I0489493

AT WHAT COST? EXAMINING THE SOCIAL COST OF CARBON

JOINT HEARING

BEFORE THE

SUBCOMMITTEE ON ENVIRONMENT &
SUBCOMMITTEE ON OVERSIGHT

COMMITTEE ON SCIENCE, SPACE, AND
TECHNOLOGY

HOUSE OF REPRESENTATIVES

ONE HUNDRED FIFTEENTH CONGRESS

FIRST SESSION

FEBRUARY 28, 2017

Serial No. 115–05

Printed for the use of the Committee on Science, Space, and Technology

Available via the World Wide Web: http://science.house.gov

U.S. GOVERNMENT PUBLISHING OFFICE

24–670PDF   WASHINGTON : 2017

For sale by the Superintendent of Documents, U.S. Government Publishing Office
Internet: bookstore.gpo.gov   Phone: toll free (866) 512–1800; DC area (202) 512–1800
Fax: (202) 512–2104   Mail: Stop IDCC, Washington, DC 20402–0001

COMMITTEE ON SCIENCE, SPACE, AND TECHNOLOGY

HON. LAMAR S. SMITH, Texas, *Chair*

FRANK D. LUCAS, Oklahoma
DANA ROHRABACHER, California
MO BROOKS, Alabama
RANDY HULTGREN, Illinois
BILL POSEY, Florida
THOMAS MASSIE, Kentucky
JIM BRIDENSTINE, Oklahoma
RANDY K. WEBER, Texas
STEPHEN KNIGHT, California
BRIAN BABIN, Texas
BARBARA COMSTOCK, Virginia
GARY PALMER, Alabama
BARRY LOUDERMILK, Georgia
RALPH LEE ABRAHAM, Louisiana
DRAIN LaHOOD, Illinois
DANIEL WEBSTER, Florida
JIM BANKS, Indiana
ANDY BIGGS, Arizona
ROGER W. MARSHALL, Kansas
NEAL P. DUNN, Florida
CLAY HIGGINS, Louisiana

EDDIE BERNICE JOHNSON, Texas
ZOE LOFGREN, California
DANIEL LIPINSKI, Illinois
SUZANNE BONAMICI, Oregon
ALAN GRAYSON, Florida
AMI BERA, California
ELIZABETH H. ESTY, Connecticut
MARC A. VEASEY, Texas
DONALD S. BEYER, JR., Virginia
JACKY ROSEN, Nevada
JERRY MCNERNEY, California
ED PERLMUTTER, Colorado
PAUL TONKO, New York
BILL FOSTER, Illinois
MARK TAKANO, California
COLLEEN HANABUSA, Hawaii
CHARLIE CRIST, Florida

SUBCOMMITTEE ON ENVIRONMENT

HON. ANDY BIGGS, Arizona, *Chair*

DANA ROHRABACHER, California
BILL POSEY, Florida
MO BROOKS, Alabama
DANIEL WEBSTER, Florida
BRIAN BABIN, Texas
GARY PALMER, Alabama
BARRY LOUDERMILK, Georgia
JIM BANKS, Indiana
CLAY HIGGINS, Louisiana
LAMAR S. SMITH, Texas

SUZANNE BONAMICI, Oregon, *Ranking Member*
COLLEEN HANABUSA, Hawaii
CHARLIE CRIST, Florida
EDDIE BERNICE JOHNSON, Texas

SUBCOMMITTEE ON OVERSIGHT

HON. DRAIN LaHOOD, Illinois, *Chair*

BILL POSEY, Florida
THOMAS MASSIE, Kentucky
GARY PALMER, Alabama
ROGER W. MARSHALL, Kansas
CLAY HIGGINS, Louisiana
LAMAR S. SMITH, Texas

DONALD S. BEYER, Jr., Virginia, *Ranking Member*
JERRY MCNERNEY, California
ED PERLMUTTER, Colorado
EDDIE BERNICE JOHNSON, Texas

CONTENTS

February 28, 2017

Opening Statements

Witnesses:

Appendix I: Answers to Post-Hearing Questions

Appendix II: Additional Material for the Record

AT WHAT COST? EXAMINING THE SOCIAL COST OF CARBON

TUESDAY, FEBRUARY 28, 2017

HOUSE OF REPRESENTATIVES,
SUBCOMMITTEE ON ENVIRONMENT AND
SUBCOMMITTEE ON OVERSIGHT,
COMMITTEE ON SCIENCE, SPACE, AND TECHNOLOGY,
Washington, D.C.

The Subcommittees met, pursuant to call, at 10:07 a.m., in Room 2318, Rayburn House Office Building, Hon. Andy Biggs [Chairman of the Subcommittee on Environment] presiding.

LAMAR S. SMITH, Texas
CHAIRMAN

EDDIE BERNICE JOHNSON, Texas
RANKING MEMBER

Congress of the United States
House of Representatives
COMMITTEE ON SCIENCE, SPACE, AND TECHNOLOGY

2321 RAYBURN HOUSE OFFICE BUILDING

WASHINGTON, DC 20515–6301

(202) 225–6371
www.science.house.gov

Subcommittees on Environment and Oversight

At What Cost? Examining the Social Cost of Carbon

Tuesday, February 28, 2017
10:00 a.m. – 11:30 a.m.
2318 Rayburn House Office Building

Witnesses

Dr. Ted Gayer, Vice President and Director of Economic Studies, Brookings Institute

Dr. Kevin Dayaratna, Senior Statistician and Research Programmer, Center for Data Analysis, The Heritage Foundation

Dr. Michael Greenstone, Milton Friedman Professor in Economics, the College, and the Harris School; Director of the Interdisciplinary Energy Policy Institute, University of Chicago; Director of Energy & Environment Lab, University of Chicago Urban Labs

Dr. Patrick Michaels, Director, Center for the Study of Science, Cato Institute

U.S. HOUSE OF REPRESENTATIVES
COMMITTEE ON SCIENCE, SPACE, AND TECHNOLOGY

HEARING CHARTER

February 24, 2017

TO: Members, Subcommittee on Environment and Subcommittee on Oversight

FROM: Majority Staff, Committee on Science, Space, and Technology

SUBJECT: Joint Subcommittee hearing: "At What Cost? Examining the Social Cost of Carbon"

The Subcommittees on Environment and Oversight will hold a joint hearing titled *At What Cost? Examining the Social Cost of Carbon* on Tuesday, February 28, 2017, at 10:00 a.m. in Room 2318 of the Rayburn House Office Building.

Hearing Purpose:

The purpose of this hearing is to examine the methods and parameters used to establish the social cost of carbon. Witnesses will discuss the models used to determine the value for the social cost of carbon and how the process can be improved.

Witness List:

- **Dr. Ted Gayer, PhD**, Vice President and Director of Economic Studies and Joseph A. Pechman Senior Fellow at Brookings Institution
- **Dr. Kevin Dayaratna, PhD**, Senior Statistician and Research Programmer, Center for Data Analysis, Institute for Economic Freedom and Opportunity at The Heritage Foundation
- **Dr. Michael Greenstone, PhD**, Milton Friedman Professor in Economics, the College, and the Harris School; Director of the interdisciplinary Energy Policy Institute at the University of Chicago and the Energy & Environment Lab at the University of Chicago Urban Labs
- **Dr. Patrick Michaels, PhD**, Director, Center for the Study of Science, Cato Institute; contributing author to United Nations Intergovernmental Panel on Climate Change (Nobel Peace Prize 2007)

Staff Contact:

For questions related to the hearing, please contact Juliya Grigoryan of the Majority Staff at 202-225-6371.

Chairman BIGGS. Good morning. The Subcommittees on Environment and Oversight will come to order.

Without objection, the Chair is authorized to declare recesses of the Subcommittee at any time.

Welcome to today's hearing entitled "At What Cost? Examining the Social Cost of Carbon." I recognize myself for five minutes for an opening statement.

Welcome to today's joint subcommittee hearing entitled "At What Cost? Examining the Social Cost of Carbon." Today, we will examine the previous Administration's determination of the social cost of carbon, or SCC, and explore why the calculated value is flawed. Energy is the bedrock of our society, and yet the SCC estimate of the previous Administration has killed jobs, limited innovation, and resulted in higher energy costs for American families, all in exchange for benefits that are negligible at best and nonexistent at worst.

The Obama Administration's Interagency Working Group, which ultimately established an enormously high SCC of $37 per ton of CO_2 emitted into the atmosphere, relied on an outdated economic model and failed to take into account the White House's own Office of Management and Budget, or OMB, guidelines for cost-benefit analysis. Quite simply, the working group used numbers that got them the results they wanted in order to advance some of the most expensive and expansive regulations ever written. In pushing forward this political agenda, the working group acted irresponsibly. It also allowed the previous Administration to implement stringent and costly regulations without a scientific basis.

As we will learn today, the SCC working group ignored two major OMB recommendations for federal agency rulemaking. First, it failed to use a seven percent discount rate, and instead relied on rates of 2.5 percent, three percent, and five percent. and, second, it ignored the guideline to report cost-benefit analysis from a domestic perspective. If nothing else is taken away from what will be a very technical hearing, I hope it will be these two very basic flaws.

The low long-term discount rate established by the previous Administration fundamentally disregards the notion that the American economy is resilient and can respond to potential future threats with technological development and innovation. As to the flaw of the previous Administration's decision to focus on CO_2 emissions from a global perspective, this approach leaves the United States footing the bill for costly regulations that are based on benefits conferred to other countries. It is simply not right for Americans to be bearing the brunt of costs when the majority of benefits will be conferred away from home.

By ignoring OMB guidelines, the current SCC models leave critical components out of the discussion. If the OMB guidelines would have been followed, the social cost of carbon would have been significantly lower.

The previous Administration disregarded scientific integrity by overestimating climate change resulting from greenhouse gas emissions. In order to push an expensive regulatory agenda, the Administration inflated the SCC to justify costly regulations in response to the allegedly terrible damage CO_2 emissions will cause in the future.

The SCC is nothing but a one-sided manipulation of parameters to fit the policy-driven agendas of the previous Administration. These alarmist tactics need to stop. Today's hearing is intended to uncover the real truth and deception behind the SCC.

America's strength emanates from our resilience and flexibility. Attempts to justify government regulations over industry innovations hinders growth and development. I look forward to working with the Trump Administration to renew faith in American ingenuity and technological development.

[The prepared statement of Chairman Biggs follows:]

COMMITTEE ON
SCIENCE, SPACE, & TECHNOLOGY
Lamar Smith, Chairman

For Immediate Release
February 28, 2017

Media Contact: Kristina Baum
(202) 225-6371

Statement of Chairman Andy Biggs (R-Ariz.)
At What Cost? Examining the Social Cost of Carbon

Chairman Biggs: Welcome to today's joint subcommittee hearing entitled "At What Cost? Examining the Social Cost of Carbon." Today we will examine the previous administration's determination of the Social Cost of Carbon, or SCC, and explore why the calculated value is flawed.

Energy is the bedrock of our society. And yet, the SCC estimate of the previous administration has killed jobs, limited innovation, and resulted in higher energy costs for American families—all in exchange for benefits that are negligible at best, and nonexistent at worst.

The Obama Administration's Interagency Working Group, which ultimately established an enormously high SCC of $37 per ton of CO_2 emitted into the atmosphere, relied on outdated economic models and failed to take into account the White House's own Office of Management and Budget, or OMB, guidelines for cost-benefit analysis.

Quite simply, the working group used numbers that got them the results they wanted in order to advance some of the most expensive and expansive regulations ever written. In pushing forward this political agenda, the working group acted irresponsibly. It also allowed the previous administration to implement stringent and costly regulations without a scientific basis

As we will learn today, the SCC working group ignored two major OMB recommendations for federal agency rulemaking. First, it failed use a 7 percent discount rate, and instead relied on rates of 2.5%, 3%, and 5%; and, second, it ignored the guideline to report cost-benefit analysis from a domestic perspective. If nothing else is taken away from what will be a very technical hearing, I hope it will be these two very basic flaws.

The low long-term discount rate established by the previous administration fundamentally disregards the notion that the American economy is resilient and can respond to potential future threats with technological development and innovation.

As to the flaw of the previous administration's decision to focus on CO_2 emissions from a global perspective, this approach leaves the U.S. footing the bill for costly regulations that are based on benefits conferred to other countries. It is simply not right for

Americans to be bearing the brunt of costs when the majority of benefits will be conferred away from home.

By ignoring OMB guidelines, the current SCC models leave critical components out of the discussion. If the OMB guidelines would have been followed, the social cost of carbon would be significantly lower.

The previous administration disregarded scientific integrity by overestimating climate change resulting from greenhouse gas emissions. In order to push an expensive regulatory agenda, the administration inflated the SCC to justify costly regulations in response to the *allegedly* terrible damage CO_2 emissions will cause in the future.

The SCC is nothing but a one-sided manipulation of parameters to fit the policy-driven agendas of the previous of the previous administration. These alarmist tactics need to stop. Today's hearing is intended to uncover the real truth and deception behind the SCC.

America's strength emanates from our resilience and flexibility. Attempts to justify government regulations over industry innovations hinders growth and development. I look forward to working with the Trump administration to renew faith in American ingenuity and technological development.

###

Chairman BIGGS. I now recognize the Ranking member of the Subcommittee on Environment, Ms. Bonamici, for an opening statement.

Ms. BONAMICI. Thank you very much, Mr. Chairman, and thank you to our witnesses for being here today.

The social cost of carbon is a metric used to value the damage caused by emitting 1 ton of carbon dioxide into the atmosphere in a year. It provides a consistent value for all federal agencies to use for their cost-benefit analysis on regulatory efforts that reduce carbon dioxide emissions.

There are some people who criticize this metric, but the Government Accountability Office and independent peer review by the National Academy of Sciences have validated it many times. Additionally, federal courts have upheld that the methodology used to develop the social cost of carbon is based on robust science and sound economic analysis. It is critical that updates to the social cost of carbon metric are based on the best available science and updated economic analysis based on peer-reviewed literature.

The Government Accountability Office has found that the methodology used to develop the social cost of carbon was based on peer-reviewed academic literature and took steps to incorporate new information as it became available. This process also provided ample opportunity for public comment on both the social cost of carbon and the regulations that use the metric in their cost-benefit analysis.

Some people suggest that regulations to reduce the emissions of carbon dioxide and other pollutants are unnecessary because climate change does not exist or human activity does not contribute to it, but simply ignoring a fact does not make it less true. The climate is warming, and we need to work now to limit the consequences for future generations. Our children and grandchildren should not inherit an environment that degrades their health and harms their future economy.

Economic growth and reducing carbon pollution are not in conflict with one another. Clean energy development allows us to continue powering our communities in ways that avoid long-term negative consequences on future generations. It also gives us the opportunity to bring new living-wage jobs into communities. In fact, the American Wind Energy Association found that the wind energy sector accounts for 3,000 jobs throughout my home State of Oregon alone. In addition to boosting Oregon's economy, wind energy generation avoided more than 1 million tons of statewide carbon dioxide emissions in 2015, and many of the wind energy jobs are in rural areas where jobs are needed.

The social cost of carbon is not a product of a single President, a single scientific study, or a single legal action. It is rooted in overwhelming scientific consensus on climate change, an effort spanning 30 years from both the executive and judicial branches of the Federal Government. These factors, coupled with a transparent development process and strong economic analysis, form the basis of this metric that has been used in at least 79 federal regulations, including fuel economy standards for vehicles, energy efficiency measures for home appliances, and regulations such as the Clean Power Plan. This metric was not invented to serve a political agenda but in fact was developed to meet a legal mandate to justify, in simple terms of dollars and cents, how the Federal Government's actions will affect Americans today and our children and grandchildren tomorrow.

I look forward to hearing how we may best continue to use the social cost of carbon in support of policies that protect our environment.

With that, I would like to again thank the witnesses for being here today, and I yield back the balance of my time.

[The prepared statement of Ms. Bonamici follows:]

<u>OPENING STATEMENT</u>
Ranking Member Suzanne Bonamici (D-OR)
of the Subcommittee on Environment

House Committee on Science, Space, & Technology
Subcommittee on Environment
Subcommittee on Oversight
"At What Cost? Examining the Social Cost of Carbon"
February 28, 2017

Thank you, Mr. Chairman. And thank you to our witnesses for being here today.

The "social cost of carbon" is a metric used to value the damage caused by emitting one ton of carbon dioxide into the atmosphere in a year. It provides a consistent value for all federal agencies to use for their cost benefit analyses on regulatory efforts that reduce carbon dioxide emissions.

There are some people who criticize this metric, but the Government Accountability Office and independent peer review by the National Academy of Sciences have validated it many times. Additionally, federal courts have upheld that the methodology used to develop the social cost of carbon is based on robust science and sound economic analysis.

It is critical that updates to the social cost of carbon metric are based on the best available science and updated economic analysis based on peer reviewed literature. The Government Accountability Office has found that the methodology used to develop the social cost of carbon was based on peer reviewed academic literature and took steps to incorporate new information as it became available. This process also provided ample opportunity for public comment on both the social cost of carbon and the regulations that use the metric in their cost benefit analyses.

Some people suggest that regulations to reduce emissions of carbon dioxide and other pollutants are unnecessary because climate change does not exist, or human activity does not contribute to it. But simply ignoring a fact does not make it any less true. The climate is warming and we need to work now to limit the consequences for future generations. Our children and grandchildren should not inherit an environment that degrades their health and harms their future economy.

Economic growth and reducing carbon pollution are not in conflict with one another. Clean energy development allows us to continue powering our communities in ways that avoid long-term negative consequences on future generations. It also gives us the opportunity to bring new living wage jobs into our communities. In fact, the American Wind Energy Association found that the wind energy sector accounts for 3,000 jobs throughout my home state of Oregon alone. In addition to boosting Oregon's economy, wind energy generation avoided more than one million tons of statewide carbon dioxide emissions in 2015.

The social cost of carbon is not a product of a single President, a single scientific study, or a single legal action. It is rooted in overwhelming scientific consensus on climate change, and efforts spanning thirty years from both the executive and judicial branches of the federal government. These factors, coupled with a transparent development process and strong economic analysis, form the basis of this metric that has been used in at least 79 federal regulations, including fuel economy standards for vehicles, energy efficiency measures for home appliances, and regulations such as the Clean Power Plan.

This metric was not invented to serve a political agenda, but in fact was developed to meet a legal mandate to justify, in simple terms of dollars and cents, how the federal government's actions will affect Americans today, and our children and grandchildren tomorrow. I look forward to hearing how we may best continue to use the social cost of carbon in support of policies that protect our environment.

With that I would like to again thank the witnesses for being here today, and I yield back the balance of my time.

Chairman BIGGS. Thank you, Ms. Bonamici.

I now recognize the Chairman of the Subcommittee on Oversight, Mr. LaHood, for his opening statement.

Mr. LAHOOD. Thank you, Chairman Biggs, and happy to be part of this hearing today with you, today's hearing titled "At What Cost? Examining the Social Cost of Carbon," and happy to have the witnesses here today as we examine the previous Administration's social cost of carbon and the shortfalls in application of this flawed process.

There is significant evidence that the previous Administration manipulated the social cost of carbon calculation to reflect significant benefits to enacting what were ultimately job-killing regulations and policies across a wide spectrum of issues. The social cost of carbon is a flawed tool used by the Obama Administration to justify a green agenda when, in reality, the prior Administration was seeking to offset its costly regulations with far-reaching implications that burden our industries and nation.

Unsurprisingly, the previous Administration ignored specific guidelines set forth by the Office of Management and Budget, OMB, and used the social cost of carbon as a vehicle to tout the economic benefits of the new environmental regulations. This is troubling and to me is not being honest with the taxpayers.

Critics take issue primarily with two aspects of the social cost of carbon methodology, specifically, the discount rate used and the domestic versus global benefits claimed. Both issues I look forward to discussing in more detail with our panel of esteemed witnesses today.

I, too, take issue with the methodology but also the lack of transparency with the use and development of the social cost of carbon. Three statistical integrated assessment economic models were used to develop the social cost of carbon: the FUND, the DICE, and the PAGE. Experts have concluded these three models are flawed and possess too many uncertainties to be the foundation of the benefit analysis of environmental regulations. If one were to change the assumptions these models are based on, the result will drastically differ, demonstrating malleability in the social cost of carbon calculation.

Because of these realities, last year, I was pleased to be an original cosponsor of H.R. 5668, the Transparency and Honesty in Energy Regulation Act, or THERA, introduced by my friend and colleague Evan Jenkins of West Virginia. This legislation is aimed at prohibiting the Department of Energy and the Environmental Protection Agency from considering the social cost of carbon as part of any cost-benefit analysis unless specifically authorized by law. If signed into law, the DOE and the EPA would no longer rely on manipulated and fabricated economic benefits to justify or support new job-killing environmental regulations. I look forward to working with Congressman Jenkins again on this issue this Congress. It appears that the social cost of carbon is nothing but a political tool lacking scientific integrity and transparency conceived and utilized by an Administration pushing a green agenda to the detriment of the American taxpayers. Perhaps a better measurement of the social cost of carbon is not the net damages that result from a one-metric-ton increase in carbon dioxide emissions in a given year but the damage inflicted on domestic industries, including manufacturers in my district like Caterpillar, by the environmental regulations justified by this flawed calculation.

I would like to thank our witnesses for being here today to discuss this important matter. In addition, I look forward working with the Trump Administration to reverse the damage caused by the Obama Administration as it relates to this issue.

With that, I yield back to the Chair.

[The prepared statement of Mr. LaHood follows:]

COMMITTEE ON
SCIENCE, SPACE, & TECHNOLOGY
Lamar Smith, Chairman

For Immediate Release
February 28, 2017

Media Contact: Kristina Baum
(202) 225-6371

Statement of Chairman Darin LaHood (R-Ill.)
At What Cost? Examining the Social Cost of Carbon

Chairman LaHood: Welcome to today's joint subcommittee hearing examining the previous administration's Social Cost of Carbon, the shortfalls and application of this flawed process. The previous administration manipulated the Social Cost of Carbon calculation to reflect significant benefits to enacting what were ultimately job-killing regulations and policies across a wide spectrum of issues. The Social Cost of Carbon is a flawed, tool used by the Obama Administration to justify a green agenda. When in reality, the prior Administration was seeking to offset its costly regulations with far reaching implications that burden our industries and nation.

Unsurprisingly the previous administration ignored specific guidelines set forth by the Office of Management and Budget (OMB) and used the Social Cost of Carbon as a vehicle to tout the economic benefits of new environmental regulations. This is troubling and to me is not being honest with the taxpayers.

Critics take issue primarily with two aspects of the Social Cost of Carbon methodology. Specifically, the discount rate used and the domestic versus global benefits claimed. Both issues I look forward to discussing in more detail with our panel of esteemed witnesses today.

I, too, take issue with the methodology but also the lack of transparency with the use and development of the Social Cost of Carbon. Three statistical integrated assessment economic models were used to develop the Social Cost of Carbon. The FUND (Climate Framework for Uncertainty, Negotiation and Distribution), the DICE (Dynamic Integrated Climate-Economy), and the PAGE (Policy Analysis of the Greenhouse Effect). Experts have concluded these three models are flawed and possess too many uncertainties to be the foundation of the benefit analysis of environmental regulations. If one were to change the reasonable assumptions these models are based on, the result will drastically differ, demonstrating malleability in the Social Cost of Carbon calculation.

Because of these realities, last year I was pleased to cosponsor H.R. 5668, Transparency and Honesty in Energy Regulation Act. Or THERA. This legislation is aimed at prohibiting the Department of Energy and the Environmental Protection Agency from considering the Social Cost of Carbon as part of any cost benefit analysis unless specifically authorized by law. If passed the DOE and EPA would no longer rely on manipulated and fabricated economic benefits to justify or support new job-killing environmental regulations.

The Social Cost of Carbon is nothing but a political tool lacking scientific integrity and transparency conceived and utilized by an administration pushing a green agenda to the detriment of the American taxpayers. Perhaps a better measurement of the Social Cost of Carbon is *not* the net damages that result from a 1-metric ton increase in carbon dioxide emissions in a given year but the damage inflicted on domestic industries by the environmental regulations justified by this flawed calculation.

I would like to thank our witnesses for being here today to discuss this important matter. In addition, I look forward working with the Trump administration to reverse the damage caused by the Obama Administration. With that, I yield back to the chair.

###

Chairman BIGGS. Thank you, Mr. LaHood.

I now recognize the Ranking Member of the Subcommittee on Oversight, Mr. Beyer, for his opening statement.

Mr. BEYER. Thank you, Mr. Chairman, and thank you, all the witnesses, for being here.

You know, the social cost of carbon is a complex metric which our witness Dr. Greenstone has described as the most important number you've never heard of. Assessing and addressing the impact of climate change on current and future generations is critical. It seems already that in just the first few minutes of this hearing we see a dramatic difference between the short-term emphasis on job creation, which is important, and a long-term emphasis on protecting our planet.

The social cost of carbon permits the government to help quantify the future economic damages as a result of carbon pollution that contributes to climate change and global warming. This metric didn't materialize out of thin and dirty air. It took a federal judge to mandate its use during the Bush Administration based on a law passed when Ronald Reagan was President.

In 2009, the Obama Administration convened an interagency effort to formalize a consistent value for it. This was not a political tool. This was an attempt to protect our environment.

We'll hear today that this development process was transparent, it was open to public comment, it's been validated over the years, and, much like our climate, it's not static and it changes over time in response to updated inputs. And although its use has been challenged in the courts recently, the courts have upheld the methodology used to obtain this estimate as proper based on real science and appropriate economic models.

As a Minnesota administrative law judge determined last April, the preponderance of evidence supports the fact that federal social cost of carbon is reasonable and the best available measure to determine the environmental cost of CO_2. I'm pretty certain we won't hear any of that today from the majority members and their witnesses. Instead, we'll hear the same arguments made against climate regulations that we've heard before. And sadly, those anti-science arguments both ignore the abundant scientific evidence that shows that climate change exists, that fossil fuel production is its main contributor, and will also admonish virtually any responsible regulatory mechanism to help protect our nation's citizens from the environmental, economic, and public health harm that results from climate change's global impacts.

These individuals will argue already that social cost of carbon is outdated, inaccurate, and not a proper regulatory mechanism for addressing climate change. We've heard these arguments before. In fact, in 1982 the tobacco company R.J. Reynolds produced internal talking points about the social cost of smoking when Congressman Henry Waxman was holding hearings regarding the harm to the public's health from cigarette smoking. At the time, Representative Waxman said the annual smoking-related cost in lost productivity was $25.8 billion and $13.6 billion in annual medical costs. R.J. Reynolds said, quote, that "attempts to establish a dollar value of so-called cost of smoking are ill-founded and unreliable," end quote.

More than one decade later, in 1994, the tobacco company Philip Morris was producing glossy brochures to combat the growing evidence revolving around the harm of cigarette smoking. One was titled, quote, "Debunking the Social Cost of Smoking," and an internal memo from Philip Morris said simply, "Philip Morris does not believe that smoking has been shown to pose any social cost on society."

So we're going to hear similar arguments today on the social cost of carbon emissions from fossil fuels, their impact on climate change. These arguments resonate loudly with the new Trump Administration, but they contradict the economic analysis and scientific evidence that supports the use of social cost of carbon.

In a much-publicized recent memo, Dr.—or Thomas Pyle, the head of Trump's Department of Energy transition team, stated, quote, "If the social cost of carbon were subjected to the latest science, it would certainly be much lower than what the Obama Administration has been using," end quote. And he suggested ending the use of it in federal rulemaking. The memo went on to describe plans to withdraw from the Paris Climate Agreement, eliminate the Clean Power Plan, increase federal oil and natural gas leasing, lift the moratorium on coal leasing—in other words, more and more and more fossil fuels at greater cost to the environment.

Mr. Chairman, climate change is real. Scientific evidence across the world, we think we are the only country in the world that doesn't—that has any internal disagreement about climate change. And as members of the Science Committee, we should be leading the fight to protect our nation against its impacts, and I hope my colleagues will be persuaded by the weight of evidence. The evidence becomes ever clearer with every passing day. And we will work together to promote policies that protect our future generations.

Mr. Chair, I yield back.

[The prepared statement of Mr. Beyer follows:]

OPENING STATEMENT
Ranking Member Don Beyer (D-VA)
of the Subcommittee on Oversight

House Committee on Science, Space, & Technology
Subcommittee on Environment
Subcommittee on Oversight
"At What Cost? Examining the Social Cost of Carbon"
February 28, 2017

Thank you, Mr. Chairman.

The social cost of carbon (SCC) is a complex metric which our witness, Dr. Greenstone, has described as "the most important number you have never heard of." Assessing and addressing the impact of climate change on current and future generations is critical. The social cost of carbon permits the government to help quantify the future economic damages caused as a result of carbon emissions that contribute to climate change and global warming.

This metric did not simply materialize out of thin, and dirty, air. It took a federal court judge to mandate its use during the Bush Administration. In 2009, the Obama administration convened an interagency effort to formalize a consistent value for it. We will hear today that this development process was transparent, has been open to public comment, has been validated over the years and, much like our climate, is not static and changes over time in response to updated inputs.

Although its use has been challenged in the courts recently, the courts have upheld the methodology used to obtain this estimate as proper, based on real science and appropriate economic models. As a Minnesota administrative law judge determined last April, a preponderance of the evidence supports the fact that the "Federal Social Cost of Caron is reasonable and the best available measure to determine the environmental cost of CO2 [Carbon Dioxide emissions] ...."

I am pretty certain we won't hear any of that from some of the Majority Members and their witnesses today. Instead, we will hear the same arguments made against climate regulations that we have heard before. Sadly, those anti-science arguments that both ignore the abundant scientific evidence that has shown climate change exists, and that fossil fuel production is the main contributor, also admonish virtually any responsible regulatory mechanisms to help protect our nation's citizens from the environmental, economic and public health harm resulting from climate change's global impacts. These individuals will argue that the Social Cost of Carbon is outdated, inaccurate, and not a proper regulatory mechanism for addressing climate change.

Four decades ago, we heard almost identical arguments in a different context. In 1982 the tobacco company RJ Reynolds produced internal talking points about the social cost of smoking when former Congressman Henry Waxman was holding hearings regarding the harm to the public's health from cigarette smoking. At the time, Rep. Waxman said the annual smoking related costs in lost productivity was $25.8 billion and $13.6 billion in annual medical costs. RJ Reynolds said that "attempts to estimate a dollar value of so-called costs of smoking are ill-founded and unreliable." More than one decade later, in 1994, the tobacco company Philip Morris was producing glossy brochures to combat the growing evidence revolving around the harm of cigarette smoking. One was titled: "Debunking the 'Social Costs' of Smoking," and an internal memo from Philip Morris said simply, "Philip Morris does not believe that smoking has been shown to impose any social cost on society."

We will hear similar arguments today regarding the social cost of carbon emissions from fossil fuels and their impact on climate change. Those arguments resonate loudly with the new Trump Administration, but contradict the economic analysis and scientific evidence that supports the use of the social cost of carbon. In a much publicized recent memo, Thomas Pyle, the head of Trump's Department of Energy (DOE) transition team, stated that, "if the [social cost of carbon] were subjected to the latest science, it would certainly be much lower than what the Obama administration has been using" and he suggested ending the use of it in federal rulemakings. The memo went on to describe plans to withdraw from the Paris Climate Agreement, eliminating the Clean Power Plan, increasing federal oil and natural gas leasing, and lifting the moratorium on coal leasing. In other words, more fossil fuels, including the dirtiest of them all - coal, at greater cost to the environment.

Mr. Chairman, climate change is real, scientific evidence supports this reality, and as Members of the Science Committee we should be leading the fight to protect our Nation against its impacts. I hope my colleagues will be persuaded by the weight of evidence and work with me to promote policies that protect our future generations, not unduly burden them by promoting actions and policies that we know are harmful to their health, the environment and the economy.

Thank you. I yield back.

Chairman BIGGS. Thank you, Mr. Beyer.

I now recognize the Chairman of the Full Committee, Mr. Smith, for his opening statement. Chairman Smith.

Chairman SMITH. Thank you, Chairman Biggs. And congratulations on becoming the Chairman of the Environment Subcommittee. I look forward to helping you restrain the EPA's out- of-control regulatory agenda.

The EPA, along with other federal agencies, often bases their regulations on models and science not familiar to most Americans. Americans are led to believe that the EPA's regulations are based on the best science available. Unfortunately, this committee has discovered that that is not the case.

The EPA's track record does not inspire trust. For example, the EPA routinely relies on nondisclosed scientific studies to justify its regulations, but how can Americans believe an agency that isn't being open and honest?

Another little-known component of environmental regulations is the social cost of carbon. The EPA attempts to put a price on a ton of carbon emitted into the atmosphere. This term is in many of the EPA's regulations. However, like many of the Agency's determinations, it is often based on a one-sided political agenda.

Many factors contribute to the value of the social cost of carbon. While multiple models are used to determine a value for carbon, the ones frequently used in regulations assume only a worst-case scenario for climate change impacts. Similar to climate models, which predict worst-case scenarios and are repeatedly proved wrong, the social cost of carbon used by federal agencies is also flawed.

The federal government should not include faulty calculations to justify costly regulations. Examples would be the Clean Power Plan and standards used by the Department of Energy. Instead, it should eliminate the use of the social cost of carbon until a credible value can be calculated.

Rushing to use unreliable calculations, such as the social cost of carbon, to justify a regulation is irresponsible and misleading. For instance, the EPA's Clean Power Plan would cost billions of dollars every year in return for a minimal benefit on the environment. In fact, the regulation would reduce global temperatures by only 0.03 degrees Celsius and limit sea level rise by only the width of three sheets of paper.

One of the many components used to justify this rule is the social cost of carbon. This flawed value desperately attempts to justify the Agency's alarmist reasoning for support of the Clean Power Plan and other climate regulations. Agencies should rely on sound science, not flawed data. The fact that different models for the social cost of carbon exist and all have different values is a testament to how uncertain the science behind the value really is. For example, the social cost of carbon ranges from negative values to $37 per ton, which is the estimate used by government agencies under the Obama Administration. Before the EPA includes this value in rulemakings, the Agency should reassess how it is determined.

Americans deserve credible science, not regulations based on data that is suspect and calculated to justify the EPA's climate agenda. Sound science and actual data should lead the way, not politically calculated social costs.

Thank you, Mr. Chairman. I'll yield back.

[The prepared statement of Chairman Smith follows:]

COMMITTEE ON
SCIENCE, SPACE, & TECHNOLOGY
Lamar Smith, Chairman

For Immediate Release
February 28, 2017

Media Contact: Kristina Baum
(202) 225-6371

Statement of Chairman Lamar Smith (R-Texas)
At What Cost? Examining the Social Cost of Carbon

Chairman Smith: Thank you Chairman Biggs and congratulations on becoming Chairman of the Environment Subcommittee. I look forward to helping you restrain the EPA's out-of-control regulatory agenda.

The EPA, along with other federal agencies, often bases their regulations on models and science not familiar to most Americans.

Americans are led to believe that the EPA's regulations are based on the best science available. Unfortunately, this Committee has uncovered that this is not the case.

The EPA's track record does not inspire trust. For example, the EPA routinely relies on non-disclosed scientific studies to justify its regulations. How can Americans believe an agency that isn't being open and honest?

Another little known component of environmental regulations is the social cost of carbon. The EPA attempts to put a price on a ton of carbon emitted into the atmosphere.

This term is in many of the EPA's regulations. However, like many of the agency's determinations, it is often based on a one-sided political agenda.

Many factors contribute to the value of the social cost of carbon.

While multiple models are used to determine a value for carbon, the ones frequently used in regulations assume only a worst case scenario for climate change impacts.

Similar to climate models, which predict worst case scenarios and are repeatedly proved wrong, the social cost of carbon used by federal agencies is also flawed.

The federal government should not include faulty calculations to justify costly regulations, such as the Clean Power Plan and standards used by the Department of Energy. Instead, it should eliminate the use of the social cost of carbon until a credible value can be calculated.

Rushing to use unreliable calculations, such as the social cost of carbon, to justify a regulation is irresponsible and misleading.

23

For instance, the EPA's Clean Power Plan would cost billions of dollars every year in return for a minimal benefit on the environment. In fact, the regulation would reduce global temperatures by only 0.03 degrees Celsius and limit sea level rise by only the width of three sheets of paper.

One of the many components used to justify this rule is the social cost of carbon. This flawed value desperately attempts to justify the agency's alarmist reasoning for support of the Clean Power Plan and other climate regulations.

Agencies should rely on sound science, not flawed data. The fact that different models for the social cost of carbon exist and all have different values is a testament to how uncertain the science behind the value really is.

The social cost of carbon ranges from negative values to $37 per ton, which is the estimate used by government agencies under the Obama administration.

Before the EPA includes this value in rulemakings, the agency should reassess how it is modeled and valued.

Americans deserve credible science, not regulations based on data that is suspect and calculated to justify the EPA's climate agenda. Sound science and actual data should lead the way, not politically calculated social costs.

###

24

Chairman BIGGS. Thank you, Chairman Smith.

Let me introduce our witnesses. Our first witness today is Dr. Ted Gayer, Vice President and Director of Economic Studies and Joseph A. Pechman Senior Fellow at the Brookings Institute. Dr. Gayer received his bachelor's degree in mathematics and economics from Emory, and his master's degree and Ph.D. in economics from Duke University.

Our next witness is Dr. Kevin Dayaratna, Senior Statistician and Research Programmer at the Heritage Foundation's Institute for Economic Freedom and Opportunities Center for Data Analysis.

Dr. Dayaratna received his bachelor's degree in applied mathematics and mathematical physics from the University of California at Berkeley, his master's degrees in mathematical statistics and business management from the University of Maryland, and his Ph.D. in mathematical statistics from the University of Maryland.

Our third witness today is Dr. Michael Greenstone, Milton Friedman Professor in Economics, the College, and the Harris School; Director of the Interdisciplinary Energy Policy Institute at the University of Chicago; and Director of the Energy and Environment Lab at University of Chicago Urban Labs. Dr. Greenstone received his bachelor's degree in economics at Swarthmore College and his Ph.D. in economics from Princeton.

Our final witness today will be Dr. Patrick Michaels, Director of the Center for the Study of Science at the Cato Institute and contributing author to United Nations' Intergovernmental Panel on Climate Change, which was awarded the Nobel Peace Prize. Dr. Michaels received his bachelor's and master's degrees in biological sciences and plant ecology from the University of Chicago and his Ph.D. in ecological climatology from the University of Wisconsin at Madison.

I now recognize Dr. Gayer for five minutes to present his testimony.

TESTIMONY OF DR. TED GAYER, PHD, VICE PRESIDENT AND DIRECTOR OF ECONOMIC STUDIES AND JOSEPH A. PECHMAN SENIOR FELLOW AT BROOKINGS INSTITUTION

Dr. GAYER. Chairs Biggs, LaHood, and Smith; and Ranking Members Bonamici and Beyer; and members of the subcommittees, I appreciate the opportunity to appear here today to discuss the social cost of carbon.

The social cost of carbon is a dollar estimate of the damages caused by a 1-ton increase in greenhouse gas emissions in a given year. It is a conceptually valid and important consideration when devising policies and treaties to address climate change, yet estimating the value of the social cost of carbon is an enormously complex and uncertain exercise.

In 2009, the Obama Administration established an Interagency Working Group to develop a range of estimates for the social cost of carbon subsequently used by agencies to evaluate federal regulations. My focus is on the specific question of whether the social cost of carbon should account for the global or the domestic harm of a ton of greenhouse gas.

In a world in which the United States and all the other major emitters of greenhouse gases adopted a coordinated set of policies to address climate change, then a global measure would be appropriate since greenhouse gases contribute to damages around the world no matter where they occur.

But we don't live in such a world. Instead, in the United States we have opted for a suite of regulatory policies ranging from subsidizing lower carbon energy sources; mandating energy efficiency levels in buildings, vehicles, and household appliances; requiring transportation fuels to contain minimum volumes of different renewable fuels, and restricting emissions from electric utilities.

Given the diversity of regulations directed at climate change, it is useful and important for the agencies to coordinate on a single measure of climate benefits. But the question is whether they should report and consider the benefits to U.S. citizens or to the world. The Interagency Working Group opted for a global measure. I believe the exclusive focus on a global measure runs counter to standard benefit cost practice in which only the benefits within the political jurisdiction bearing the cost of the policy are considers. It also seems at odds with the express intent of longstanding Execu- tive orders and of authorizing statutes. For example, the main reg- ulatory guidance document that has been in place for 20 years is Executive Order 12866, which makes clear that the appropriate reference point for analyzing federal regulatory policy—policies is the U.S. citizenry, not the world.

Similarly, when enacting the Clean Air Act, Congress stated that its purpose was to, quote, "protect and enhance the quality of the nation's air resources so as to promote the public health and welfare and productive capacity of its population," end quote, which again suggests a focus on domestic benefits.

The global measure is 4 to 14 times greater than the estimated domestic measure, which is significant. For example, for its proposed regulations for existing power plants, the EPA estimated climate benefits amounting to $30 billion in 2030. However, the estimated domestic climate benefits would have only amounted to $2 to $7 billion, which is less than EPA's estimated compliance cost for the rule.

I believe that adopting a global measure for the benefits of a domestic policy would be justified if U.S. actions led to complete reciprocity from other countries. The question is whether efforts by the United States to regulate greenhouse gases might spur reciprocity by other countries to do so as well, generating domestic benefits that are 4 to 14 times as great as the direct domestic benefits to the U.S.-only policy. This is doubtful since the agency regulations taken under existing U.S. laws such as the Clean Air Act are not tantamount to treaty commitments that can establish a formal basis for other countries matching the efforts undertaken domestically.

By using the global social cost of carbon, the agencies are claiming that their rules provide benefits that in fact largely accrue to foreign citizens. Of course, many Americans are altruistic and care about the welfare of people beyond our borders, but foreign aid decisions should be made openly and not hidden in an obscure metric used in domestic rulemaking.

A global measure of the social cost of carbon is appropriate if the intent is to use it to support the development of a global system of reducing greenhouse gases, such as through a worldwide carbon tax. I favor a carbon tax for the United States that replaces regulations and relies on border tax adjustments to incentivize other major emitters to follow suit. But absent such an approach, for domestic agencies considering domestic regulations in which the costs are incurred domestically, a global measure deviates from standard practice and requires more scrutiny and justification than it has received to date. At the very least, agencies should report the expected domestic benefits and only separately and transparently report the expected foreign benefits of their actions informed by concrete evidence of reciprocity expected from other countries.

Thank you.

[The prepared statement of Dr. Gayer follows:]

One Page Summary of Ted Gayer's Testimony

My testimony addresses whether estimates of the social cost of carbon should consider the global or only the domestic costs of greenhouse gas emissions. The key points are:

- A global measure of the social cost of carbon is appropriate if the intent is to use it to support the development of a global system of reducing greenhouse gases, in which U.S actions are completely reciprocated.

- Absent such an approach, for domestic agencies considering domestic regulations, in which the costs are incurred domestically, a global measure deviates from standard practice and seems at odds with the intent of long-standing executive orders and authorizing statutes.

- The global measure is 4 to 14 times greater than the estimated domestic measure, which is significant. By exclusively using the global social cost of carbon, agencies are claiming that their rules provide benefits that in fact largely accrue to foreign citizens.

- Use of a global measure requires much more scrutiny and justification than it has received to date. At the very least, agencies should report the expected domestic benefits and only separately and transparently report the expected foreign benefits of their actions, informed by concrete evidence of reciprocity expected from other countries.

Testimony of Ted Gayer
Senior Fellow, Brookings Institution
Before the Subcommittee on Environment and the Subcommittee on Oversight,
Committee on Science, Space, & Technology,
U.S. House of Representatives,
February 28, 2017

Chairs Biggs and LaHood, Ranking Members Bonamici and Beyer, and Members of the Subcommittees on Environment and on Oversight, I appreciate the opportunity to appear here today to discuss the social cost of carbon.

The social cost of carbon is an estimate of the monetized damages caused by a one-ton increase in greenhouse gas emissions in a given year. It is a conceptually valid and important consideration when devising policies and treaties to address climate change.

Yet estimating the value of the social cost of carbon is an enormously complex and uncertain exercise. It requires understanding the effect of a ton of a greenhouse gas on global temperatures; the effect of temperature change on agricultural yields, human health, flood risk, and myriad other harms to the ecosystem; monetizing these various damages into dollar terms; and determining how much to balance harm to future generations against the interests of the current generation. In 2009, the U.S. government established an interagency working group, composed of scientific and economic experts from the White House and a number of agencies, to develop a range of estimates for the social cost of carbon, subsequently used by agencies to evaluate federal regulations.

My focus is on the specific question of whether the social cost of carbon should account for the global or the domestic harm of a ton of a greenhouse gas.[1] In a world in which the United States and all the other major emitters of greenhouse gases adopted a coordinated set of policies to address climate change, a global measure would be appropriate, since greenhouse gases contribute to damages around the world no matter where they occur.

But we don't live in such a world. Instead, in the U.S. we have opted for a suite of regulatory policies, ranging from subsidizing lower-carbon energy sources, mandating energy efficiency levels in buildings, vehicles, and household appliances, requiring transportation fuels to contain minimum volumes of different renewable fuels, and restricting emissions from electric utilities. Given the diversity of regulations directed at climate change, it is useful and important for the agencies to coordinate on a single measure for the social cost of carbon. But the question is

---

[1] Much of my testimony is drawn from work I have done with W. Kip Viscusi and shorter pieces co-authored with Susan Dudley, Art Fraas, John Graham, Randall Lutter, Jason F. Shogren, and W. Kip Viscusi. I have submitted some of these as part of my written statement.

whether they should report and consider the climate benefits to U.S. citizens or to the world. The interagency working group opted for a global measure, which has since been the basis for considering the benefits associated with all climate-related regulations.

I believe that the exclusive focus on a global measure runs counter to standard benefit–cost practice, in which only the benefits within the political jurisdiction bearing the cost of the policy are considered. It also seems at odds with the expressed intent of long-standing executive orders and of authorizing statutes. For example, the main regulatory guidance document that has been in place for over 20 years is Executive Order 12866, which makes clear that the appropriate reference point for analyzing federal regulatory policies is the U.S citizenry, not the world. And a subsequent guidance document by the Office of Management and Budget (known as Circular A-4) maintained an emphasis on domestic benefits. Similarly, when enacting the Clean Air Act, Congress stated that its purpose was to "protect and enhance the quality of the Nation's air resources so as to promote the public health and welfare and productive capacity of its population," which again suggests a focus on domestic benefits. Similar language is found in other authorizing statutes for environmental regulations.

The difference between global and domestic benefits of greenhouse gas regulations is significant, as the global measure is 4 to 14 times greater than the estimated domestic measure. For example, for its proposed regulations for existing power plants, the EPA estimated climate benefits amounting to $30 billion in 2030. However, the estimated domestic climate benefits only amount to $2-$7 billion, which is less than EPA's estimated compliance costs for the rule of $7.3 billion. The use of a global social cost of carbon to estimate benefits means that agencies will adopt regulations that could cost Americans more than they receive in climate-related benefits. This approach could be especially problematic if U.S. actions simply shift emissions overseas.

I believe that adopting a global measure for the benefits of a domestic policy would be justified if U.S. actions led to complete reciprocation from other countries. The question is whether efforts by the United States to regulate greenhouse gases might spur reciprocity by other countries to do so as well, generating domestic benefits that are 4 to 14 times as great as the direct domestic benefits to the U.S.-only policy. This is doubtful, since the regulations taken under existing U.S. laws, such as the Clean Air Act, are not tantamount to treaty commitments that can establish a formal basis for other countries matching the efforts undertaken domestically.

By using the global social cost of carbon, the agencies are claiming that their rules—which impose substantial domestic costs—provide benefits that in fact largely accrue to foreign citizens. Of course, many Americans are altruistic and care about the welfare of people beyond our borders. But foreign aid decisions should be made openly, not hidden in an obscure metric used in rulemaking.

A global measure of the social cost of carbon is appropriate if the intent is to use it to support the development of a global system of reducing greenhouse gas emissions, such as through a worldwide carbon tax. I favor a carbon tax for the U.S. that replaces regulations and relies on border-tax adjustments to incentivize other major emitters to follow suit. But, absent such an approach, for domestic agencies considering domestic regulations, in which the costs are incurred domestically, a global measure deviates from standard practice and requires much more scrutiny and justification than it has received to date. At the very least, agencies should report the expected domestic benefits and only separately and transparently report the expected foreign benefits of their actions, informed by evidence of concrete reciprocation expected from other countries. Thank you.

Ted Gayer's Short Summary Biography

Ted Gayer is the vice president and director of the Economic Studies program and the Joseph A. Pechman Senior Fellow at the Brookings Institution. He conducts research on a variety of economic issues, focusing particularly on public finance, environmental and energy economics, housing, and regulatory policy.

Prior to joining the Brookings Institution in September 2009, he was associate professor of public policy at Georgetown University. From 2007 to 2008, he was deputy assistant secretary for Economic Policy at the Department of the Treasury. While at Treasury, he worked primarily on housing and credit market policies, as well as on energy and environmental issues, health care, Social Security and Medicare.

From 2003 to 2004, he was a senior economist at the President's Council of Economic Advisers, where he worked on environmental and energy policies. From 2006 to 2007, he was a visiting fellow at the Public Policy Institute of California, and from 2004 to 2006 he was a visiting scholar at the American Enterprise Institute.

Chairman BIGGS. Thank you, Dr. Gayer.

I now recognize Dr. Dayaratna for five minutes to present his testimony.

TESTIMONY OF DR. KEVIN DAYARATNA, PHD, SENIOR STATISTICIAN AND RESEARCH PROGRAMMER, CENTER FOR DATA ANALYSIS, INSTITUTE FOR ECONOMIC FREEDOM AND OPPORTUNITY AT THE HERITAGE FOUNDATION

Dr. DAYARATNA. Chairman Biggs, Ranking Member Bonamici, and other Members of the Subcommittees, thank you for the opportunity to testify about the social cost of carbon.

My name is Kevin Dayaratna. I'm the Senior Statistician and Research Programmer at the Heritage Foundation. The views I express in this testimony are my own and should not be construed as representing any official position of the Heritage Foundation.

One of the primary metrics that the previous Administration had used to justify agenda regarding energy policy—justify regulatory—its regulatory agenda regarding energy policy is the social cost of carbon, which is defined as the economic damages associated with a metric ton of carbon dioxide emissions summed across a particular time horizon.

There are three primary statistical models that the Obama Administration's Interagency Working Group had used to estimate the SCC, the DC. model, the FUND model, and the PAGE model. My colleagues and I have used the DC. and FUND models, testing their sensitivity to a variety of important assumptions. Our work, published both at Heritage, as well as in the peer-reviewed literature, as repeatedly illustrated that while these models might be interesting for academic exercises, they can be readily manipulated by regulators and bureaucrats.

In particular, as with any statistical model, they are dependent on various assumptions. I'd like to discuss three assumptions regularly manipulated to achieve predetermined outcomes: the choice of a discount rate, a time horizon, and the specification of an equilibrium climate sensitivity distribution.

The first easily manipulated assumption is the discount rate. In this type of cost-benefit analysis, the discount rate should reflect the rate of return on generally achievable alternative investments. The IWG had run these models using 2.5, 3, and five percent discount rates despite the fact that OMB guidance in circular A–4 had specifically stipulated that a seven percent discount rate be used as well.

At Heritage, we re-estimated these models using a seven percent discount rate and noticed drastic reductions to the SCC. In 2020, for example, according to our recent analysis of the DC. model published in the peer-reviewed journal Climate Change Economics, under a three percent discount rate, the SCC is estimated to cost $37.79 per ton, while under a seven percent discount rate, it is estimated to be $5.87, an 84 percent reduction. The higher estimates previously found by the IWG can enable policymakers to justify unnecessary regulations and taxes on the economy.

The second easily manipulated assumption is the specification of a time horizon. It is close to impossible to forecast what the economy will look like decades into the future. Foolishly, these models attempt to make projections not decades but rather three centuries into the future. In my work at Heritage, I have changed the time horizon to a significantly shorter but still unrealistic time horizon of 150 years into the future. With the DC. model, we find that these results plummet by 25 percent in some instances.

The third readily manipulated variable is the model's equilibrium climate sensitivity, or ECS, distribution, quantifying the Earth's temperature response to a doubling of carbon dioxide concentration. My colleague Dr. Pat Michaels will go into this in more detail, but the IWG used an ECS distribution that was published ten years ago in the journal Science. Since then, a number of newer ECS distributions have been published suggesting lower probabilities of extreme global warming.

Using the more up-to-date ECS distributions generate significantly lower estimates of the SCC. In our peer-reviewed work, we found that, as a result of updating the ECS distributions, the results drop by as much as 197 percent under some circumstances. Inflated estimates of climate sensitivity drive up the SCC, which can become manifested in unnecessary regulations.

Finally, the unexplored issue here is are there any benefits associated with carbon dioxide emissions? The answer is surprisingly yes. The FUND model actually allows for negative SCC, meaning a positive outcome. In fact, under some assumptions, there are actual substantial probabilities of negative SCC, meaning increased CO_2 fertilization, leading to increased agriculture and forestry yields.

Moreover, if one were to take the IWG's interpretation of these models seriously and implement the associated regulations, there would be significant damage to the economy. In particular, our analysis finds that by 2035 the country would experience an average employment shortfall of 400,000 lost jobs, a marked increase in electricity prices, and an aggregate $2.5 trillion loss in GDP.

Our analysis using the model for the assessment of greenhouse gas-induced climate change has found that these devastating impacts would be accompanied by insignificant changes and less than 2/10 of a degree Celsius in temperature mitigation and less than 2 centimeters of sea level rise reduction.

In conclusion, the SCC is a broken tool for regulatory policy and taking it seriously would provide significant harm and little environmental benefit.

Thank you for your attention, and I look forward to your questions.

[The prepared statement of Mr. Dayaratna follows:]

At What Cost? Examining the Social Cost of Carbon
Kevin D. Dayaratna, Ph.D.
Senior Statistician and Research Programmer - The Heritage Foundation

1. The Social Cost of Carbon (SCC) is a tool used by policymakers to quantify the economic damages associated with carbon dioxide emissions. In my work at The Heritage Foundation, we have rigorously examined two of the three models that the Obama Administration's Interagency Working Group (IWG) used to estimate the SCC. This work has been published both at The Heritage Foundation as well as the peer reviewed literature.

2. The models are extremely sensitive to very reasonable changes to assumptions. As a result, these models can be manipulated to produce a wide range of costs.

3. The models are based on projections 300 years into the future. It is difficult to envision what the country would look like decades, let alone centuries into the future. Upon changing this time span to the less unrealistic time horizon of 150 years into the future, we found that the estimates plummet by as much as 25% in some instances.

4. The Administration's analysis of the SCC assumes an outdated climate sensitivity specification based on a paper published ten years ago in the journal *Science*. This specification is no longer defensible. We have re-estimated the SCC using more up-to date distributions and found reductions of up to nearly 200%. The use of this outdated distribution thus artificially inflates the calculated value of the SCC.

5. The Office of Management and Budget stipulated in Circular A-4 that a 7% discount rate be used as part of cost-benefit analysis. The Administration's IWG ignored this recommendation. We reran the models using a 7% discount rate and found that the SCC drops by over 75% when compared to a 3% discount rate.

6. Under a variety of assumptions, including those made by the IWG itself, one of its three predictive models shows that the SCC has a non-trivial probability of being negative. This would suggest that there are actually benefits of CO_2 emissions. Under some very reasonable assumptions, this probability (~70%) can be quite substantial.

7. The GHG regulations implied by the IWG's use of these models would result in significant damage to the economy. Our analysis finds that, by 2035, the country would experience an average employment shortfall of 400,000 lost jobs, a total loss of income over $20,000 for a family of four, a 13-20% increase in electricity prices, and an aggregate $2.5 trillion loss in GDP.

8. In addition to the above damages, these regulations would result in negligible environmental benefits (<0.2°C temperature mitigation and less than 2 cm of sea level reductions).

The Heritage Foundation

CONGRESSIONAL TESTIMONY

Methods and Parameters Used to Establish the Social Cost of Carbon

Testimony before the Subcomittee on Environment and Oversight

Committee on Science and Technology

U.S. House of Representatives

February 24, 2017

Kevin D. Dayaratna, PhD
Senior Statistician and Research Programmer
The Heritage Foundation

Chairman Biggs, Ranking Member Bonamici, and other Members of the subcommittee, thank you for the opportunity to testify about the social cost of carbon. My name is Kevin Dayaratna. I am the Senior Statistician and Research Programmer at The Heritage Foundation. The views I express in this testimony are my own and should not be construed as representing any official position of The Heritage Foundation.

For much of the past decade, the federal government has strived to expand regulations across the energy sector of the economy. One of the primary justifications for doing so has been the social cost of carbon (SCC), which is defined as the economic damages associated with a metric ton of carbon dioxide (CO_2) emissions summed across a particular time horizon.[1]

The Models

There are three primary statistical models that the Interagency Working Group (IWG) has used to estimate the SCC—the Dynamic Integrated Climate-Economy (DICE) model, the Framework for Uncertainty, Negotiation and Distribution (FUND) model, and the Policy Analysis of the Greenhouse Effect (PAGE) model.[2] Over the last several years at

The Heritage Foundation, my colleagues and I have used the DICE and FUND models, testing their sensitivity to a variety of important assumptions. Our research, published as Heritage Foundation publications, in the peer-reviewed literature, and discussed in my prior congressional testimony, has repeatedly illustrated that although these models might be interesting academic exercises, they are extremely sensitive to very reasonable changes to assumptions.[3] These models can thus be manipulated by user-selected assumptions, and are therefore not suitable for guiding regulatory policy.

These models are estimated by Monte Carlo simulation. The general idea behind Monte Carlo simulation is that since some aspects of the models are random, the models are repeatedly estimated to generate a spectrum of probable outcomes. As a result of principles in probability theory, repeated estimation for a sufficient amount of time provides a reasonable characterization of the SCC's distributional properties.

As with any statistical model, however, these models are grounded by assumptions. In our work, my colleagues and I have rigorously examined three important assumptions: the choice of a discount rate, a time horizon, and the specification of an equilibrium climate sensitivity distribution.

38

CONGRESSIONAL TESTIMONY

TABLE 1

DICE Model Average SCC – Baseline, End Year 2300

Year	Discount Rate - 2.50%	Discount Rate - 3%	Discount Rate - 5%	Discount Rate - 7%
2010	$46.58	$30.04	$8.81	$4.02
2020	$56.92	$37.79	$12.10	$5.87
2030	$66.53	$45.15	$15.33	$7.70
2040	$76.96	$53.26	$19.02	$9.85
2050	$87.70	$61.72	$23.06	$12.25

SOURCE: Kevin Dayaratna, Ross McKitrick, and David Kreutzer, "Empirically-Constrained Climate Sensitivity and the Social Cost of Carbon," *Climate Change Economics*.

☎ heritage.org

TABLE 2

FUND Model Average SCC – Baseline, End Year 2300

Year	Discount Rate - 2.50%	Discount Rate - 3%	Discount Rate - 5%	Discount Rate - 7%
2010	$29.69	$16.98	$1.87	-$0.53
2020	$32.90	$19.33	$2.54	-$0.37
2030	$36.16	$21.78	$3.31	-$0.13
2040	$39.53	$24.36	$4.21	$0.19
2050	$42.98	$27.06	$5.25	$0.63

SOURCE: Kevin Dayaratna, Ross McKitrick, and David Kreutzer, "Empirically-Constrained Climate Sensitivity and the Social Cost of Carbon," *Climate Change Economics*.

☎ heritage.org

Discount Rate

As people prefer benefits sooner rather than later and costs later rather than sooner, discount rates enable us to normalize inequalities regarding long-term investments. The Environmental Protection Agency (EPA) has run these models using 2.5 percent, 3 percent, and 5 percent discount rates despite the fact that the Office of Management and Budget guidance in Circular A-4 has specifically stipulated that a 7 percent discount rate be used as well.[4] In my research, we re-estimated these models using a 7 percent discount rate in a variety of publications, holding constant all other assumptions made by the IWG. Tables 1 and 2 are our results recently accepted for publication in the peer-reviewed journal *Climate Change Economics*.

As we can see, the SCC estimates are drastically reduced under the use of a 7 percent discount rate. In fact, under the FUND model, the estimates are negative, suggesting that there are actually benefits to CO_2 emissions. These changes in the discount rate can cause the SCC to drop by as much as 80 percent or more.

Time Horizon

It is essentially impossible to forecast technological change decades, let alone centuries, into the future. Regardless, however, these SCC models are based on projections 300 years into the future. In my work at Heritage, I have changed this time horizon to the significantly less, albeit still unrealistic,

TABLE 3

DICE Model Average SCC - End Year 2150

Year	Discount Rate - 2.50%	Discount Rate - 3%	Discount Rate - 5%	Discount Rate - 7%
2010	$36.78	$26.01	$8.66	$4.01
2020	$44.41	$32.38	$11.85	$5.85
2030	$50.82	$38.00	$14.92	$7.67
2040	$57.17	$43.79	$18.36	$9.79
2050	$62.81	$49.20	$22.00	$12.13

SOURCE: Kevin D. Dayaratna and David W. Kreutzer, "Loaded DICE: An EPA Model Not Ready for the Big Game," Heritage Foundation Backgrounder No. 2860, November 21, 2013, http://www.heritage.org/environment/report/loaded-dice-epa-model-not-ready-the-big-game.

🐘 heritage.org

time horizon of 150 years into the future, and we obtained the following results for the DICE model in our work published in 2013 (see Table 3).[5]

Clearly, the SCC estimates drop substantially as a result of changing the end year (in some cases by over 25 percent).

Equilibrium Climate Sensitivity (ECS) Distribution

These models of course take into account assumptions regarding the planet's climate sensitivity. The real question, however, is the degree of accuracy statistical models have at doing so. Dr. John Christy testified in both 2013 and 2016 regarding the efficacy of climate change projections and juxtaposed them against reality. In his testimony, Christy exposed the sheer inadequacy of the Intergovernmental Panel on Climate Change's (IPCC's) models in forecasting global temperatures.[6]

The climate specification used in estimating the SCC is that of an equilibrium climate sensitivity (ECS) distribution. These distributions probabilistically quantify the earth's temperature response to a doubling of CO_2 concentrations. The ECS distribution used by the IWG is based on a paper published in the journal *Science* ten years ago by Gerard Roe and Marcia Baker. This non-empirical distribution, calibrated by the IWG based on assumptions that the group decided on climate change in conjunction with IPCC recommendations, has been deemed to be "no longer scientifically defensible."[7] Since then, a variety of newer and more up-to-date distributions have been suggested in the peer-reviewed

literature. Many of these distributions, in fact, suggest lower probabilities of extreme global warming in response to CO_2 concentrations. Below are a few such distributions:[8]

The area under the curve between two temperature points depicts the probability that the Earth's temperature will increase between those amounts in response to a doubling of CO_2 concentrations. Thus, the area under the curve from 4 degrees Celsius (C) onwards (known as a "tail probability") provides the probability that the Earth's temperature will warm by more than 4 degrees C in response to a doubling of CO_2 concentrations. Note that the more up-to-date ECS distributions (Otto et al., 2013; Lewis, 2013; Lewis and Curry, 2015; see chart on p. 7) have significantly lower tail probabilities (5 to 700 times lower regarding temperature increases above 4 degrees C) than the outdated Roe-Baker (2007) distribution used by the IWG. In our research published in *Climate Change Economics*, we re-estimated the SCC having used these more up-to-date ECS distributions and obtained the following results (see Tables 4 and 5).[9]

Again, we notice drastically lower estimates of the SCC using these more up-to-date ECS distributions. These results are not surprising—the IWG's estimates of the SCC were based on outdated assumptions that overstated the probabilities of extreme global warming, which artificially inflated their estimates of the SCC.

Negativity

When people talk about the social cost of carbon, they tend to think of damages. Not all of these models,

TABLE 4

DICE Model Average SCC – ECS Distribution Updated in Accordance with Lewis and Curry (2015), End Year 2300

Year	Discount Rate - 2.50%	Discount Rate - 3%	Discount Rate - 5%	Discount Rate - 7%
2010	$23.62	$15.62	$5.03	$2.48
2020	$28.92	$19.66	$6.86	$3.57
2030	$33.95	$23.56	$8.67	$4.65
2040	$39.47	$27.88	$10.74	$5.91
2050	$45.34	$32.51	$13.03	$7.32

SOURCE: Kevin Dayaratna, Ross McKitrick, and David Kreutzer, "Empirically-Constrained Climate Sensitivity and the Social Cost of Carbon," *Climate Change Economics.*

☎ heritage.org

TABLE 5

FUND Model Average SCC – ECS Distribution Updated in Accordance with Lewis and Curry (2015), End Year 2300

Year	Discount Rate - 2.50%	Discount Rate - 3%	Discount Rate - 5%	Discount Rate - 7%
2010	$5.25	$2.78	-$0.65	-$1.12
2020	$5.86	$3.33	-$0.47	-$1.10
2030	$6.45	$3.90	-$0.19	-$1.01
2040	$7.02	$4.49	-$0.18	-$0.82
2050	$7.53	$5.09	$0.64	-$0.53

SOURCE: Kevin Dayaratna, Ross McKitrick, and David Kreutzer, "Empirically-Constrained Climate Sensitivity and the Social Cost of Carbon," *Climate Change Economics.*

☎ heritage.org

however, suggest that there are always damages associated with CO_2 emissions. The FUND model, in fact, allows for the SCC to be negative based on feedback mechanisms due to CO_2 emissions. In my research at The Heritage Foundation, we actually calculated the probability of a negative SCC under a variety of assumptions. Below are some of our results published at Heritage as well as in the peer-reviewed journal *Climate Change Economics* (see Tables 6, 7, 8, and 9).[10]

As the above statistics illustrate, under a very reasonable set of assumptions, the SCC is overwhelmingly likely to be negative, which would suggest the government should, in fact, subsidize (not limit) CO_2 emissions. Of course, I by no means use these results to suggest that the government should actually subsidize CO_2 emissions, but rather to illustrate the extreme sensitivity of these models to reasonable changes to assumptions and can thus be quite easily fixed by policymakers.

Economic Growth

In 2013, Professor Robert Pindyck of MIT has summarized many of the issues associated with integrated assessment modeling:

TABLE 6

FUND Model Probability of Negative SCC – ECS Distribution Based on Outdated Roe–Baker (2007) Distribution, End Year 2300

Year	Discount Rate - 2.50%	Discount Rate - 3%	Discount Rate - 5%	Discount Rate - 7%
2010	0.087	0.121	0.372	0.642
2020	0.084	0.115	0.344	0.601
2030	0.08	0.108	0.312	0.555
2040	0.075	0.101	0.282	0.507
2050	0.071	0.093	0.251	0.455

SOURCE: Kevin Dayaratna, Ross McKitrick, and David Kreutzer, "Empirically-Constrained Climate Sensitivity and the Social Cost of Carbon," *Climate Change Economics*.

☎ heritage.org

TABLE 7

FUND Model Probability of Negative SCC – ECS Distribution Updated in Accordance with Otto et al. (2013), End Year 2300

Year	Discount Rate - 2.50%	Discount Rate - 3%	Discount Rate - 5%	Discount Rate - 7%
2010	0.278	0.321	0.529	0.701
2020	0.268	0.306	0.496	0.661
2030	0.255	0.291	0.461	0.619
2040	0.244	0.274	0.425	0.571
2050	0.228	0.256	0.386	0.517

SOURCE: Kevin D. Dayaratna and David W. Kreutzer, "Unfounded FUND: Yet Another EPA Model Not Ready for the Big Game," Heritage Foundation Backgrounder No. 2897, April 29, 2014, http://www.heritage.org/environment/report/unfounded-fund-yet-another-epa-model-not-ready-the-big-game.

☎ heritage.org

Given all of the effort that has gone into developing and using IAMs, have they helped us resolve the wide disagreement over the size of the SCC? Is the U.S. government estimate of $21 per ton (or the updated estimate of $33 per ton) a reliable or otherwise useful number? What have these IAMs (and related models) told us? I will argue that the answer is very little. As I discuss below, the models are so deeply flawed as to be close to useless as tools for policy analysis. Worse yet, precision that is simply illusory, and can be highly misleading.

...[A]n IAM-based analysis suggests a level of knowledge and precision that is nonexistent, and allows the modeler to obtain almost any desired result because key inputs can be chosen arbitrarily.[11]

What is interesting is the relationship these models have amongst SCC, temperature, and economic growth. Intuitively, one would believe that if there are indeed so-called social costs of CO_2 emissions, then they would result literal economic damages (that would be manifested in gross domestic product

42

TABLE 8

FUND Model Probability of Negative SCC – ECS Distribution Updated in Accordance with Lewis (2013), End Year 2300

Year	Discount Rate - 2.50%	Discount Rate - 3%	Discount Rate - 5%	Discount Rate - 7%
2010	0.39	0.431	0.598	0.722
2020	0.375	0.411	0.565	0.685
2030	0.361	0.392	0.53	0.645
2040	0.344	0.371	0.491	0.598
2050	0.326	0.349	0.449	0.545

SOURCE: Kevin D. Dayaratna and David W. Kreutzer, "Unfounded FUND: Yet Another EPA Model Not Ready for the Big Game," Heritage Foundation Backgrounder No. 2897, April 29, 2014, http://www.heritage.org/environment/report/unfounded-fund-yet-another-epa-model-not-ready-the-big-game.

☎ heritage.org

TABLE 9

FUND Model Probability of Negative SCC – ECS Distribution Updated in Accordance with Lewis (2013), End Year 2300

Year	Discount Rate - 2.50%	Discount Rate - 3%	Discount Rate - 5%	Discount Rate - 7%
2010	0.416	0.45	0.601	0.73
2020	0.402	0.432	0.57	0.69
2030	0.388	0.414	0.536	0.646
2040	0.371	0.394	0.496	0.597
2050	0.354	0.372	0.456	0.542

SOURCE: Kevin Dayaratna, Ross McKitrick, and David Kreutzer, "Empirically-Constrained Climate Sensitivity and the Social Cost of Carbon," *Climate Change Economics*.

☎ heritage.org

(GDP)) in the long run. These models, however, operate in a manner that is precisely the contrary. The models estimate the SCC after averaging simulations run across five different economic-growth scenarios. The plots on pages 8 and 9 provide temperature and GDP projections based on the DICE model from our 2013 analysis:[12]

The wealthiest society depicted by IMAGE has the greatest SCC estimate of the economic-growth scenarios, but only a modest amount of temperature change. As a result, the implication would be to sacrifice more economically for not necessarily more

global warming. These figures clearly demonstrate the sheer absurdity associated with the DICE model.

The Social Costs of Methane and Nitrous Oxide

The EPA has also proposed similar models to quantify the social costs of methane (SCM) and nitrous oxide emissions (SCN20). We performed a similar analysis to what is outlined above, and also noticed that these models are quite sensitive to assumptions. In particular, changes to the discount rate as well as the ECS distribution can result in reductions of the

Outdated Roe-Baker (2007) and More Recent ECS Distributions

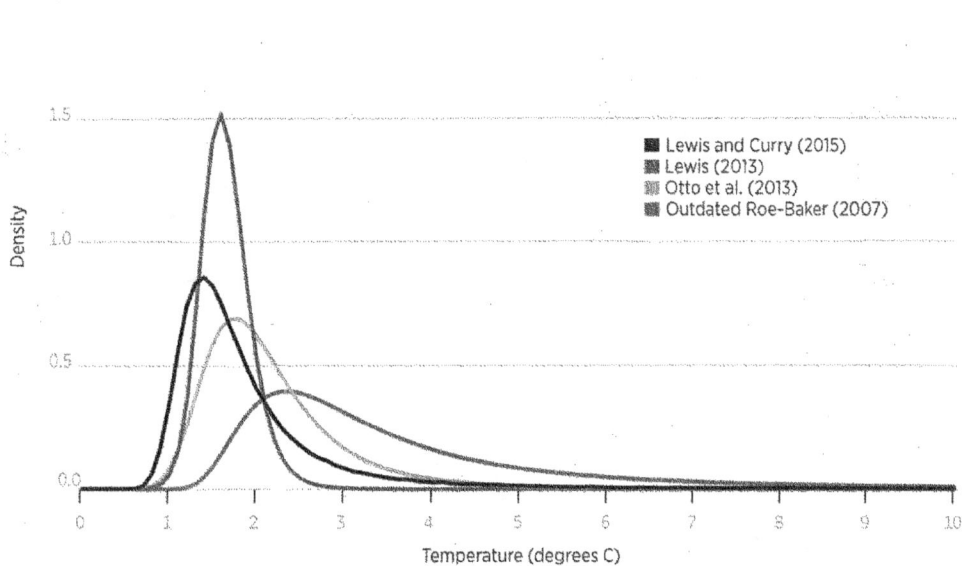

SOURCES: Gerard H. Roe and Marcia B. Baker, "Why Is Climate Sensitivity So Unpredictable?" *Science*, Vol. 318, No. 5850 (October 26, 2007), pp. 629–632; Nicholas Lewis, "An Objective Bayesian Improved Approach for Applying Optimal Fingerprint Techniques to Estimate Climate Sensitivity," *Journal of Climate*, Vol. 26, No. 19 (October 2013), pp. 7414–7429; and Alexander Otto et al., "Energy Budget Constraints on Climate Response," *Nature Geoscience*, Vol. 6, No. 6 (June 2013), pp. 415–416; Nicholas Lewis and Judith A. Curry, "The Implications for Climate Sensitivity of AR5 Forcing and Heat Uptake Estimates," *Climate Dynamics*, Vol. 45, Issue 3, pp 1009-1923. http://link.springer.com/article/10.1007/s00382-014-2342-y (accessed February 27, 2017).

SCM and SCN20 by up to 80 percent. Thus, these models, like the SCC models, can also be effortlessly manipulated by user-selected assumptions.[13]

Negligible Environmental Benefits

Given the sensitivity of these models to quite reasonable changes to assumptions, there is no reason to take them seriously for the purposes of policymaking. Regardless, we estimated the environmental impact of the associated regulations using the Model for the Assessment of Greenhouse Gas Induced Climate Change, and we simulated the environmental impact of eliminating greenhouse gas emissions from the United States completely. Even assuming a climate far more sensitive than the indefensible specifications made by the IWG in its analysis, simulation results indicate that if all carbon dioxide, methane, and nitrous oxide emissions were to be eliminated

from the United States completely, the result in terms of temperature reductions would be less than 0.2 degrees C, 0.03 degrees C, and 0.02 degrees C, respectively. These temperature reductions would also be accompanied by miniscule changes in sea level rise (less than 2 centimeter reduction).[14]

Economic Consequences

On top of the aforementioned negligible environmental benefits, our research at Heritage has demonstrated that if the greenhouse gas regulations associated with these integrated models were actually implemented, the country would suffer disastrous economic consequences. Most notably, by 2035, the country would experience an average employment shortfall of 400,000 lost jobs, a total loss of income of over $20,000 for a family of four, a 13 percent to 20 percent increase in electricity prices, and an aggregate $2.5

44

Temperature Change (DICE)

FUTURE TEMPERATURE CHANGES (IN DEGREES CELSIUS)

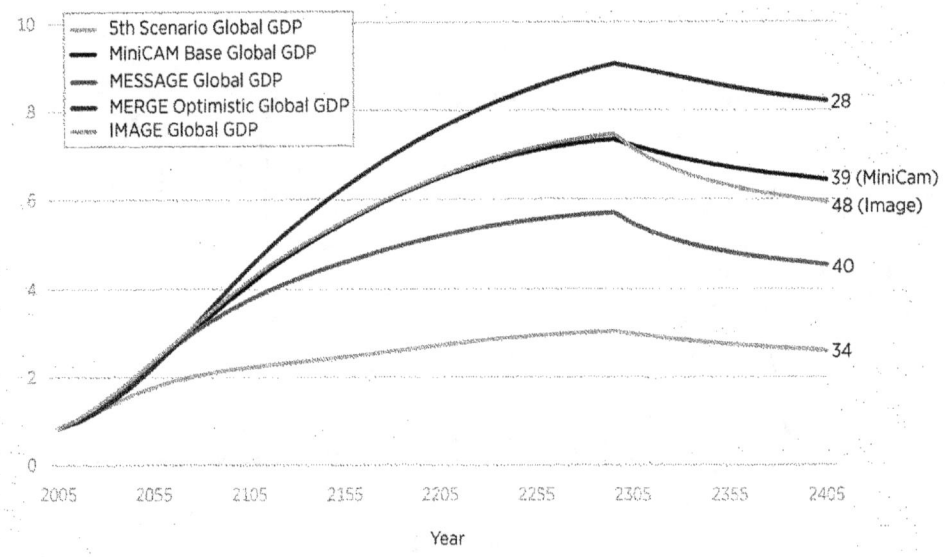

NOTE: The 2020 value of the SCC (in $2007) produced by the DICE model (assuming a 3% discount rate) is included on the right-hand side of the figure.
SOURCE: Patrick J. Michaels, "An Analysis of the Obama Administration's Social Cost of Carbon," testimony before the Committee on Natural Resources, U.S. House of Representatives, July 22, 2015, https://www.cato.org/publications/testimony/analysis-obama-administrations-social-cost-carbon (accessed February 27, 2017).

trillion loss in GDP. We have published other research in previous years, and they have also illustrated similar devastating consequences. On the other hand, taking advantage of the vast carbon-related sources of energy, such as shale oil and gas, will have essentially the opposite effect on the country—growing the economy, increasing household incomes, and adding hundreds of thousands of jobs for years to come.[15]

Criticisms

Critics may argue that the SCC has been underreported by the IWG. Much of this research, however, still suffers from many of the flaws discussed above. Furthermore, there are also questions regarding the legitimacy of the research that these studies are based on. Moore and Diaz (2015), for example, base their research on statistically insignificant results

regarding the relationship between climate change and economic growth.

Altogether, there have in fact been nearly a thousand different estimates of the SCC, with results literally all across the map. Havernek et al. (2015) provides a nice summary of these estimates and finds that the IWG's reported results are higher than what the overall peer-reviewed literatures suggest.[16]

Conclusions

The SCC (as well as the SCM and SCN20) are based on statistical models that are extremely sensitive to important assumptions incorporated within the models. The climate sensitivity specifications the models make are outdated. Moreover, the damage functions that the estimates are based on are essentially arbitrary with limited empirical

Global GDP (DICE)

FUTURE GLOBAL GROSS DOMESTIC PRODUCT (IN BILLIONS OF DOLLARS).

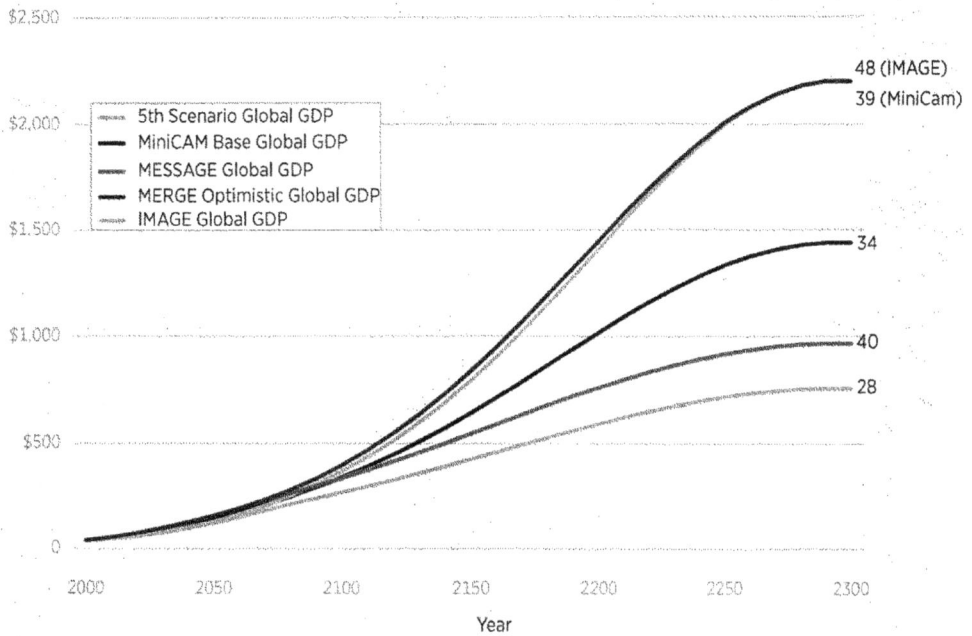

NOTE: The 2020 value of the SCC (in $2007) produced by the DICE model (assuming a 3% discount rate) is included on the right-hand side of the figure.

SOURCE: Patrick J. Michaels, "An Analysis of the Obama Administration's Social Cost of Carbon," testimony before the Committee on Natural Resources, U.S. House of Representatives, July 22, 2015, https://www.cato.org/publications/testimony/analysis-obama-administrations-social-cost-carbon (accessed February 27, 2017).

☎ heritage.org

justification. Even if one were to take their results seriously, their use would result in significant economic damages with little benefit to reducing global temperatures. As a result, these models, although they may be interesting academic exercises, are far too unreliable for use in energy policy rulemaking and can be quite easily manipulated by user-selected assumptions. We thus urge policymakers to refrain from using them in devising regulatory policy.

****

The Heritage Foundation is a public policy, research, and educational organization recognized as exempt under section 501(c)(3) of the Internal Revenue Code. It is privately supported and receives no funds from any government at any level, nor does it perform any government or other contract work.

The Heritage Foundation is the most broadly supported think tank in the United States. During 2016, it had hundreds of thousands of individual, foundation, and corporate supporters representing every state in the U.S. Its 2016 income came from the following sources:

- Individuals 75.3%

- Foundations 20.3%

- Corporations 1.8%

- Program revenue and other income 2.6%

The top five corporate givers provided The Heritage Foundation with 1.0% of its 2016 income. The Heritage Foundation's books are audited annually by the national accounting firm of RSM US, LLP.

Endnotes

1. The official definition of the social cost of carbon is the economic damages per metric ton of CO_2 emissions, and is discussed further in U.S. Environmental Protection Agency, "The Social Cost of Carbon," http://www.epa.gov/climatechange/EPAactivities/economics/scc.html (accessed September 14, 2013).

2. For the DICE model, see William D. Nordhaus, "RICE and DICE Models of Economics of Climate Change," Yale University, November 2006, http://www.econ.yale.edu/~nordhaus/homepage/dicemodels.htm (accessed November 6, 2013). For the FUND model, see "FUND—Climate Framework for Uncertainty, Negotiation and Distribution," http://www.fund-model.org/ (accessed November 6, 2013). For the PAGE model, see Climate CoLab, "PAGE," http://climatecolab.org/resources/-/wiki/Main/PAGE (accessed November 6, 2013). See also

   U.S. Interagency Working Group on Social Cost of Greenhouse Gases, "Technical Support Document: Technical Update of the Social Cost of Carbon for Regulatory Impact Analysis Under Executive Order 12866," May 2013, revised November 2013, https://www.epa.gov/sites/production/files/2016-12/documents/sc_co2_tsd_august_2016.pdf (accessed February 23, 2017); U.S. Interagency Working Group on Social Cost of Greenhouse Gases, "Addendum to Technical Support Document on Social Cost of Carbon for Regulatory Impact Analyses under Executive Order 12866: Application of Methodology to Estimate the Social Cost of Methane and the Social Cost of Nitrous Oxide," August 2016, https://www.epa.gov/sites/production/files/2016-12/documents/addendum_to_sc-ghg_tsd_august_2016.pdf (accessed February 23, 2017); and U.S. Interagency Working Group on Social Cost of Greenhouse Gases, "2010 Technical Support Document: Social Cost of Carbon for Regulatory Impact Analysis under Executive Order 12866," February 2010, https://www.epa.gov/sites/production/files/2016-12/documents/scc_tsd_2010.pdf (accessed February 23, 2017).

3. Kevin D. Dayaratna and David W. Kreutzer, "Unfounded FUND: Yet Another EPA Model Not Ready for the Big Game," Heritage Foundation *Backgrounder* No. 2897, April 29, 2014, http://www.heritage.org/research/reports/2014/04/unfounded-fund-yet-another-epa-model-not-ready-for-the-big-game; Kevin D. Dayaratna and David W. Kreutzer, "Loaded DICE: An EPA Model Not Ready for the Big Game," Heritage Foundation *Backgrounder* No. 2860, November 21, 2013, http://www.heritage.org/research/reports/2013/11/loaded-dice-an-epa-model-not-ready-for-the-big-game; and Kevin D. Dayaratna, and David Kreutzer, "Environment: Social Cost of Carbon Statistical Modeling Is Smoke and Mirrors," *Natural Gas & Electricity*, Vol. 30, No. 12 (2014), pp. 7–11;

   Kevin Dayaratna, Ross McKitrick, and David Kreutzer, "Empirically-Constrained Climate Sensitivity and the Social Cost of Carbon," *Climate Change Economics*, Forthcoming, accepted February 17, 2017; and Kevin D Dayarantna, "An Analysis of the Obama Administration's Social Cost of Carbon," testimony before the Committee on Natural Resources, U.S. House of Representatives, July 23, 2015.

4. Office of Management and Budget, "Circular A-4," White House, https://obamawhitehouse.archives.gov/omb/circulars_a004_a-4/ (accessed February 22, 2017), and Paul C. "Chip" Knappenberger, "An Example of the Abuse of the Social Cost of Carbon," Cato-at-Liberty, http://www.cato.org/blog/example-abuse-social-cost-carbon (accessed September 14, 2013).

5. Dayaratna and Kreutzer, "Loaded DICE: An EPA Model Not Ready for the Big Game."

6. John R. Christy, testimony before the Committee on Science, Space & Technology, U.S. House of Representatives, February 2, 2016, and John R. Christy, "A Factual Look at the Relationship Between Climate and Weather," testimony before the Subcommittee on Environment, Committee on Natural Resources, U.S. House of Representatives, December 11, 2013.

7. Patrick J. Michaels, "An Analysis of the Obama Administration's Social Cost of Carbon," testimony before the Committee on Natural Resources, U.S. House of Representatives, July 22, 2015, https://www.cato.org/publications/testimony/analysis-obama-administrations-social-cost-carbon (accessed February 23, 2017).

8. Gerard H. Roe and Marcia B. Baker, "Why Is Climate Sensitivity So Unpredictable?" *Science*, Vol. 318, No. 5850 (October 26, 2007), pp. 629–632; Nicholas Lewis, "An Objective Bayesian Improved Approach for Applying Optimal Fingerprint Techniques to Estimate Climate Sensitivity," *Journal of Climate*, Vol. 26, No. 19 (October 2013), pp. 7414–7429; and Alexander Otto et al., "Energy Budget Constraints on Climate Response," *Nature Geoscience*, Vol. 6, No. 6 (June 2013), pp. 415–416; Nicholas Lewis and Judith A. Curry, "The Implications for Climate Sensitivity of AR5 Forcing and Heat Uptake Estimates," *Climate Dynamics*, Vol. 45, No. 3 (September 25, 2014), pp. 1009-1923, http://link.springer.com/article/10.1007/s00382-014-2342-y (accessed February 23, 2017); and U.S. Interagency Working Group on Social Cost of Greenhouse Gases, "2010 Technical Support Document: Social Cost of Carbon for Regulatory Impact Analysis under Executive Order 12866," February 2010, https://www.epa.gov/sites/production/files/2016-12/documents/scc_tsd_2010.pdf (accessed February 23, 2017).

9. Dayaratna, McKitrick, and Kreutzer, "Empirically-Constrained Climate Sensitivity and the Social Cost of Carbon."

10. Dayaratna and Kreutzer, "Unfounded FUND: Yet Another EPA Model Not Ready for the Big Game," and Dayaratna, McKitrick, and Kreutzer, "Empirically-Constrained Climate Sensitivity and the Social Cost of Carbon."

11. R. S. Pindyck, "Climate Change Policy: What Do the Models Tell Us?" *Journal of Economic Literature*, Vol. 51, No. 3 (2013), pp. 860–872, and Patrick J. Michaels, "An Analysis of the Obama Administration's Social Cost of Carbon," testimony before the Committee on Natural Resources, U.S. House of Representatives, July 22, 2015, https://www.cato.org/publications/testimony/analysis-obama-administrations-social-cost-carbon (accessed February 23, 2017).

12. Dayaratna and Kreutzer, "Loaded DICE: An EPA Model Not Ready for the Big Game."

13. Kevin Dayaratna and Nicolas Loris, "Rolling the DICE on Environmental Regulations: A Close Look at the Social Cost of Methane and Nitrous Oxide." Heritage Foundation *Backgrounder* No. 3184, January 19, 2017, http://www.heritage.org/energy-economics/report/rolling-the-dice-environmental-regulations-close-look-the-social-cost.

48

---

14. University Corporation for Atmospheric Research, "MAGICC/SCENGEN," http://www.cgd.ucar.edu/cas/wigley/magicc/ (accessed January 9, 2017); Dayaratna and Loris, "Rolling the DICE on Environmental Regulations: A Close Look at the Social Cost of Methane and Nitrous Oxide"; Kevin Dayaratna, Nicolas Loris, and David Kreutzer, "Consequences of Paris Protocol: Devastating Economic Costs, Essentially Zero Environmental Benefits," Heritage Foundation *Backgrounder* No. 3080, April 13, 2016, http://www.heritage.org/environment/report/consequences-paris-protocol-devastating-economic-costs-essentially-zero; and Paul Knappenberger, "Analysis of US and State-by-State Carbon Dioxide Emissions and Potential "Savings" in Future Global Temperature and Global Sea Level Rise", Science & Public Policy Institute Original Paper, April 2013, http://scienceandpublicpolicy.org/images/stories/papers/originals/state_by_state.pdf (accessed February 27, 2017).

15. Kevin D. Dayaratna, Nicolas D. Loris, and David W. Kreutzer, "The Obama Administration's Climate Agenda: Will Hit Manufacturing Hard," Heritage Foundation *Backgrounder* No. 2990, November 13, 2014, http://www.heritage.org/research/reports/2014/11/the-obama-administrations-climate-agenda-underestimated-costs-and-exaggerated-benefits; Kevin D. Dayaratna, Nicolas D. Loris, and David W. Kreutzer, "The Obama Administration's Climate Agenda: Underestimated Costs and Exaggerated Benefits," Heritage Foundation *Backgrounder* No. 2975, November 13, 2014, http://www.heritage.org/research/reports/2014/11/the-obama-administrations-climate-agenda-underestimated-costs-and-exaggerated-benefits; Nicholas D. Loris, Kevin Dayaratna, and David W. Kreutzer, "EPA Power Plant Regulations: A Backdoor Energy Tax," Heritage Foundation *Backgrounder* No. 2863, December 5, 2013, http://www.heritage.org/research/reports/2013/12/epa-power-plant-regulations-a-backdoor-energy-tax; David W. Kreutzer, Nicholas D. Loris, and Kevin Dayaratna, "Cost of a Climate Policy: The Economic Impact of Obama's Climate Action Plan," Heritage Foundation *Issue Brief* No. 3978, June 27, 2013, http://www.heritage.org/research/reports/2013/06/climate-policy-economic-impact-and-cost-of-obama-s-climate-action-plan; David W. Kreutzer and Kevin Dayaratna, "Boxer-Sanders Carbon Tax: Economic Impact," Heritage Foundation *Issue Brief* No. 3905, April 11, 2013, http://www.heritage.org/research/reports/2013/04/boxer-sanders-carbon-tax-economic-impact; Dayaratna, Loris, and Kreutzer, "Consequences of Paris Protocol: Devastating Economic Costs, Essentially Zero Environmental Benefits"; and Kevin Dayaratna, David Kreutzer, and Nicholas Loris, "Time to Unlock America's Vast Oil and Gas Resources," Heritage Foundation *Backgrounder* No. 3148, September 1, 2016, http://www.heritage.org/environment/report/time-unlock-americas-vast-oil-and-gas-resources.

16. M. Dell et al., "Temperature Shocks and Economic Growth: Evidence from the Last Half Century," *American Economic Journal: Macroeconomics*, Vol. 4 (2012), pp. 66–95, http://dx.doi.org/10.1257/mac.4.3.66 (accessed February 23, 2017); F. C. Moore, and D. B. Diaz, "Temperature Impacts on Economic Growth Warrant Stringent Mitigation Policy, *Nature Climate Change*, 2015, http://doi:10.1038/nclimate2481 (accessed February 23, 2017); J. C. J. M. van den Bergh and W. J. W. Botzen, " A Lower Bound to the Social Cost of CO2 Emissions," *Nature Climate Change*, Vol. 4 (2014), pp. 253–258, http://doi:10.1038/NCLIMATE2135 (accessed February 23, 2017); Patrick J. Michaels, "An Analysis of the Obama Administration's Social Cost of Carbon," testimony before the Committee on Natural Resources, U.S. House of Representatives, July 22, 2015, https://www.cato.org/publications/testimony/analysis-obama-administrations-social-cost-carbon (accessed February 23, 2017).

Kevin D. Dayaratna, Ph.D.
Senior Statistician and Research Programmer
The Heritage Foundation

Kevin D. Dayaratna is Senior Statistician and Research Programmer in The Heritage Foundation's Center for Data Analysis (CDA). An applied statistician, he has researched and published on the use of high-powered statistical models in public policy, medical outcomes, business, economics, and even professional sports.

Dayaratna, who joined CDA in September 2012, previously was a graduate fellow in Heritage's Center for Health Policy Studies. His fellowship paper, on comparing outcomes for Medicaid patients with those for the privately insured, was cited by the American Medical Association, the National Center for Policy Analysis, and the Galen Institute, among other groups.

Dayaratna is part of the CDA team that maintains scores of databases and statistical models to support policy research; provides confidential reviews of legislation for members of Congress and the White House; and supplies data and analysis for news organizations. Census Bureau, Internal Revenue Service, Social Security, Medicare and Department of Education are only a few of the agencies and programs included in the databases.

At CDA, Dayaratna instituted the Heritage Energy Model, derived from the Energy Information Administration's National Energy Modeling System, to quantify and help policymakers understand the long-term economic effects of energy policy proposals. He has also published extensive research on integrated assessment modeling regarding the social cost of carbon, methane, and nitrous oxide. In addition to energy modeling, has Dayaratna also works on statistical modeling regarding important climate, tax, labor, health care, welfare, and entitlement policy questions.

Dayaratna grew up in Princeton Junction, N.J. He did his undergraduate work at the University of California, Berkeley, majoring in applied mathematics with a specialty in mathematical physics. He also holds two masters degrees from the University of Maryland, one in business and management and the other in mathematical statistics. In 2014, Dayaratna completed his Ph.D. in mathematical statistics from the University of Maryland with specialties in Bayesian modeling and statistical computing. His doctoral dissertation was titled "Contributions to Bayesian Statistical Modeling in Public Policy Research."

50

Chairman BIGGS. Thank you, Dr. Dayaratna.

I now recognize Dr. Greenstone for five minutes to present his testimony.

**TESTIMONY OF DR. MICHAEL GREENSTONE, PHD,
MILTON FRIEDMAN PROFESSOR IN ECONOMICS,
THE COLLEGE, AND THE HARRIS SCHOOL;
DIRECTOR OF THE INTERDISCIPLINARY
ENERGY POLICY INSTITUTE AT THE UNIVERSITY OF CHICAGO
AND THE ENERGY & ENVIRONMENT LAB
AT THE UNIVERSITY OF CHICAGO URBAN LABS**

Dr. GREENSTONE. Thank you, Chairmen Biggs and LaHood, Ranking Members Bonamici and Beyer, and Members of the Subcommittees, for inviting me to speak today.

My name is Michael Greenstone, and I'm the Milton Friedman Professor in Economics and Director of the Energy Policy Institute at the University of Chicago.

The social cost of carbon is a monetized value of the damages from the release of an additional ton of CO_2. This means that it can be used to determine the benefits of regulations that reduce CO_2 emissions. Indeed, these benefits can then be compared to the costs that regulations impose to determine whether the regulation is beneficial or not.

In 2009, while working in the Obama Administration, Cass Sunstein and I convened and co-led an Interagency Working Group to determine a government-wide value for the social cost of carbon. Ultimately, the Interagency Working Group determined a central estimate of $21 per metric ton. That estimate has since been revised to reflect scientific advances and is now about $36.

The approach has been judged valid. Last August, the Federal Court of Appeals rejected a legal challenge to the metric. Further, the General Accounting Office has said that the working group's processes and methods reflected key principles that ensured its credibility: It used consensus-based decision-making, relied largely on existing academic literature and models; and disclosed limitations and incorporated new information by considering public comments and revising the estimates as updated research became available.

Indeed, the social cost of carbon's credibility is underscored by the fact that it has been adopted by the governments of California, Illinois, Minnesota, Maine, New York, and Washington State, not to mention Canada and Mexico.

Before concluding my testimony today, I would like to address two frequent critiques of the social cost of carbon. One such critique is that the real discount rates used—2.5, 3, and 5 percent—are too low. Why is a discount rate so important? If we choose a discount rate that is too low, then we're going to pay too much for mitigation efforts today. If instead we choose one that's too high, then we will impose higher climate damages on our children and grandchildren than we intend.

Economic theory tells us that we'll be best off if the discount rate is equal to the market interest rate from investments that match the structure of payoffs that climate mitigation provides. If we thought climate damages were likely to be imposed consistently and predictably over time, then it would be appropriate to set a discount rate equal to something like the average return for the stock market. That's about 5.3 percent over the last 50 years.

But, on the other hand, if we think climate damages could be unpredictable and that tail risk points towards major losses, then markets, markets themselves, tell us to use a lower discount rate. Consider the case of gold. Its average return is only about three percent, yet people hold it as an investment. Why is that? The reason is that it pays off dramatically during infrequent episodes of economic distress. For example, during the Great Recession, gold outperformed the stock market by 67 percent.

When one considers the possibility of large temperature changes for given increases in emissions, great sea level rise in relatively short periods of time, the possibility of physical tipping points or human responses to these changes that include mass migration, then the case for using a low discount rate to determine the social cost of carbon appears strong.

In addition to this conceptual reason to prefer low discount rates, the decline in global interest rates, so another market-based reason, since the mid-1980s provides another reason. The three percent real discount rate that has been a cornerstone of regulatory analysis since 2003 draws its justification from the fact that it was roughly equal to the real rate of return on long-term government debt at that time.

However, the world has changed. Rates are now much lower and indeed the comparable rate is now probably below two percent. Put another way, capital markets are trying to tell us to use discount rates that are lower than those currently being used to determine the social cost of carbon.

A second criticism of the social cost of carbon is that it reflects global costs from emissions, but the United States should only be concerned with domestic damages. However, this criticism misses an important point that the goal of policy is to maximize net benefits to Americans and that recognizing foreign damages is likely to increase net benefits.

Why is this the case? It's because each ton of CO_2 emitted outside the United States inflicts damages on us. Thus, we benefit when China, India, the European Union, and others reduce their emissions. It absolutely strains credibility to assume that these countries' emissions cuts would be as large as if we reverted to a social cost of carbon based only on domestic damages.

In many respects, the Paris Climate Agreement, where nearly 200 countries agreed to take action on carbon emissions, is a validation of the importance of treating climate change is a global problem.

To summarize, society needs to balance the cost to our economy of mitigating climate change with the coming climate damages. Wishing that we did not face this tradeoff will not make it go away. Ultimately, we will be better off if a social cost of carbon based on sound science, economics, and law, continues to serve as a lynchpin of regulatory policy. Thank you.

[The prepared statement of Mr. Greenstone follows:]

Statement of Michael Greenstone
Milton Friedman Professor in Economics, the College and the Harris School
University of Chicago
Director, Energy Policy Institute at the University of Chicago

To be presented to:
United States House Committee on Science, Space and Technology, Subcommittee on Environment,
Subcommittee on Oversight, hearing on "At What Cost? Examining the Social Cost of Carbon"

February 28, 2017

Thank you Chairman Biggs, Chairman LaHood, Ranking Member Bonamici, Ranking Member Beyer
and members of the Subcommittees on Environment and Oversight for inviting me to speak today.

My name is Michael Greenstone, and I am the Milton Friedman Professor in Economics and
Director of the Energy Policy Institute at the University of Chicago. My research focuses on estimating
the costs and benefits of environmental quality, with a particular emphasis on the impacts of government
regulations.

The social cost of carbon is a key metric used to assess the costs and benefits of environmental
regulations that aim to reduce greenhouse gas emissions. It is the monetary cost of the damages caused by
the release of an additional ton of carbon dioxide into the atmosphere. Simply put, it reflects the cost of
climate change—accounting for the destruction of property from storms and floods, declining agricultural
and labor productivity, elevated mortality rates, and so forth.

It is perhaps the most critical component of regulatory policy in this area because, by calculating the costs
of climate change, the social cost of carbon allows for the calculation of the monetary benefits of
regulations that reduce greenhouse gases. So, for example, a regulation that reduces carbon dioxide
emissions by 10 tons would have societal benefits of $100 if the value of the social cost of carbon were
$10. These benefits can then be compared to the costs that the regulation imposes to determine whether
the regulation is socially beneficial on net. The social cost of carbon has been used to guide the design of
about 80 regulations since its original release in 2010.

As such, I appreciate the opportunity to speak with you today about the methods and parameters used to
establish the social cost of carbon. I will make several points today that I first summarize here:

1. The courts have ruled that the federal government must both regulate greenhouse gases and
   develop an estimate of the costs of these emissions. The United States government's social
   cost of carbon is a response to these rulings. It is also a key tool in the government's reliance
   on cost-benefit analysis to guide regulatory policy, which President Reagan helped to
   institutionalize in 1981.

2. The methods and models used to determine the Social Cost of Carbon have been supported
   by the Government Accountability Office and upheld by the courts. The National Academy
   of Sciences has suggested some improvements to these methods.

3. The models used to develop the social cost of carbon are based on what was the best available
   peer-reviewed scientific and economic studies. The updates since its initial release in 2010
   reflect advances in scientific understanding.

4. The use of global damages reflects the character of the climate problem and is likely to be beneficial to the United States because it will motivate emissions cuts in other countries that benefit us. The case for using a discount rate higher than 3 percent to calculate the social cost of carbon is weak and indeed there are good reasons to choose a lower discount rate.

5. Ultimately, society needs to balance the costs to our economy of mitigating climate change today with the coming climate damages. Wishing that we did not face this trade-off will not make it go away. The social cost of carbon provides a scientifically and legally valid guidepost to help us responsibly meet this balance. Its credibility is underscored by the fact that it has been adopted by the governments of California, Illinois, Minnesota, Maine, New York, and Washington, as well as Canada and Mexico.

I. Background

The social cost of carbon builds on a long tradition that has sought to bring transparency to the regulatory process. That tradition began in 1981 when President Ronald Reagan issued an executive order institutionalizing the idea that regulatory action should be implemented only in cases when "the potential benefits to society for the regulation outweigh the potential costs to society." It sounds obvious. But this idea of applying cost-benefit analyses in the regulatory arena fundamentally altered the way in which regulations were considered. Democratic and Republican leaders have since followed President Reagan's lead, ensuring that regulations pass the cost-benefit test.

Fast forward to 2007, when the Supreme Court ruled in Massachusetts vs. U.S. EPA[1] that the U.S. Environmental Protection Agency (EPA) could not sidestep its authority to regulate the greenhouse gases that contribute to global climate change unless it could provide a scientific basis for its refusal. The EPA did the opposite, providing a scientific basis for action with the subsequent Endangerment Finding[2]. The Endangerment Finding determined that greenhouse gases—including carbon dioxide and methane, among others—threaten the public health and welfare of current and future generations.

The courts mandated that the United States regulate greenhouse gases, and the laws of the land mandated that those regulations incorporate a cost-benefit analysis. The third part of this equation was solidified in 2008, when the 9th Circuit Court of Appeals ruled[3] that the Department of Transportation needed to update its regulatory impact analysis for fuel economy rules with an estimate of the social cost of carbon. The court directed that, "while the record shows that there is a range of values, the value of carbon emissions reduction is certainly not zero."

So to review: The United States is required to regulate greenhouse gases, use a cost-benefit analysis within those regulations, and incorporate a social cost of carbon *greater than zero* into the cost-benefit analysis.

Under that landscape, the Department of Energy, the Department of Transportation and EPA began to incorporate a variety of individually developed estimates of the social cost of carbon into their regulatory analyses. These estimates were derived from academic literature and ranged from zero—which they were instructed by the court to no longer use—to $159 per metric ton of carbon dioxide emitted.

II. The Development of the U.S. Government Social Cost of Carbon and its Validation

---

[1] Massachusetts v. Environmental Protection Agency, 549 U.S. 497 (2007).
[2] Environmental Protection Agency, *Endangerment and Cause or Contribute Findings for Greenhouse Gases Under Section 202(a) of the Clean Air Act*, 40 CFR Chapter 1, Vol. 74, No. 239, 15 December, 2009.
[3] *Center for Biological Diversity v. National Highway Traffic Safety Administration*, 538 F. 3d 1172 (9th Cir. 2008)

To improve consistency in the government's use of the social cost of carbon, I, then the chief economist for President Obama's Council of Economic Advisors, along with Cass Sunstein, then the administrator of the White House Office of Information and Regulatory Affairs and now a professor at Harvard, assembled and co-led an interagency working group to determine one government-wide metric. The team consisted of the top economists, scientists and lawyers from four other offices in the Executive Office of the President and six federal agencies, including the EPA and the Departments of Agriculture, Commerce, Energy, Transportation and Treasury.

The process for developing the social cost of carbon took approximately a year and included an intense assessment of the best available peer-reviewed research, and significant debate and discussion amongst the team of climate scientists, economists, lawyers and other experts across the federal government. It also included a careful consideration of public comments on the interim values agencies had been using and an interim value determined by the interagency group. Ultimately, the interagency working group determined[4] a central estimate of $21 per metric ton. That estimate has since been revised to reflect scientific advances and is now about $36.

Both the Government Accountability Office (GAO) and the courts have judged the approach used to determine the social cost of carbon to be valid. Specifically, last August a federal court of appeals rejected a legal challenge to the social cost of carbon by a trade association of refrigerator companies. The association contended that the government lacked the legal authority to consider the social cost of carbon and that its judgments were arbitrary. The court responded that it had no doubt that Congress intended to allow consideration of the social cost of carbon and that the government's judgments were reasonable.

Further, in a 2014 report[5], the GAO said that the working group's processes and methods for developing the estimates reflected three key principles that ensured its credibility as a valid approach. First, it used consensus-based decision-making. Second, it relied largely on existing academic literature and models, including technical assistance from outside resources. Third, it disclosed limitations and incorporated new information by considering public comments and revising the estimates as updated research became available.

I'd like to elaborate further on this third point regarding public comment and the need for revisions. Since 2008, agencies have published about 80 regulatory actions for public comment in the Federal Register that use social cost of carbon estimates. The agencies received many comments on the estimates through this process, and they were discussed and considered by the working group with each update.

In fact, when the working group originally convened, it did so in part to consider public comments on the interim values that agencies had used in several rules. The working group decided to revise the estimates for the first time in 2013 after agencies received a number of public comments encouraging revisions because the models used to develop the 2010 estimates had been subsequently updated.

Then, in November 2013, in response to calls for additional transparency, the Office of Management and Budget (OMB) published a specific request for public comments on the updated social cost of carbon estimate and the methodology used. This was considered a supplement to the comments already routinely received when agencies use the social cost of carbon in specific rulemakings. In response, OMB

---

[4] Interagency Working Group on Social Cost of Carbon, United States Government, *Technical Support Document: Social Cost of Carbon for Regulatory Impact Analysis Under Executive Order 12866*, February 2010.
[5] United States Government Accountability Office, *Development of Social Cost of Carbon Estimates*, Regulatory Impact Analysis, GAO-14-663, July 2014.

received[6] about 150 substantive comments, as well as about 39,000 form letters that expressed support for the efforts to establish one government-wide metric. OMB subsequently published a detailed summary and formal response to the many thoughtful comments and, in 2015, issued an updated social cost of carbon estimate.

The need to update the social cost of carbon was driven in part by the comments received. But, it was also acknowledged as a necessity by the working group from the start. The Technical Support Document clearly states:

"It is important to emphasize that the interagency process is committed to updating these estimates as the science and economic understanding of climate change and its impacts on society improves over time. Specifically, we have set a preliminary goal of revisiting the social cost of carbon values within two years or at such time as substantially updated models become available, and to continue to support research in this area. In the meantime, we will continue to explore the issues raised in this document and consider public comments as part of the ongoing interagency process."

The working group has adhered to this founding commitment. In keeping up with the latest available science and economics, the social cost of carbon has increased as the peer review literature on climate change has advanced to uncover increases in the expected costs associated with climate change. Whether future research will lead to upward or downward adjustments, or will indicate no change, sound regulatory policy demands that the social cost of carbon reflect any advances in understanding.

Finally, governments around the world have recognized the credibility of the United States government's social cost of carbon. For example, it has been adopted by the governments of California, Illinois, Minnesota, Maine, New York, and Washington, as well as Canada and Mexico.

III.   Future Revisions and the National Academies of Sciences

To ensure that the next social cost of carbon update keeps up with the latest available science and economics, in 2015 OMB directed the National Academies of Sciences (NAS) to help in providing advice on the pros and cons of potential approaches to future updates, informed by on-going public comments and the peer-reviewed literature.

The NAS released its recommendations[7] last month after a comprehensive assessment, for which I served as a reviewer. I also testified before the NAS on ways to improve the calculation of climate damages by taking advantage of new research and data. Recognizing that our social and economic understanding of the impacts of climate change have advanced greatly since the original social cost of carbon was released seven years ago, the NAS report identifies important ways to take advantage of those improvements in our understanding. It does so by providing a new framework that would strengthen the scientific basis, provide greater transparency, and improve characterization of the uncertainties of the estimates.

As a blueprint for the future, the report makes a number of recommendations aimed at helping the process "draw more readily on expertise from the wide range of scientific disciplines relevant to estimation." Importantly, this is work my colleagues and I are currently leading as part of an interdisciplinary, inter-organizational effort to calculate hyper-localized climate damages throughout the United States and globally, an effort that would provide further depth to future estimates of the social cost of carbon.

---

[6] Shelanski, Howard and Maurice Obstfeld, "Estimating the Benefits from Carbon Dioxide Emissions Reductions," *The White House of President Barack Obama,* Archives, 2 July 2015.

[7] National Academies of Sciences, Engineering, and Medicine, *Valuing Climate Damages: Updating Estimation of the Social Cost of Carbon Dioxide,* 2017.

IV.   Discount Rates and Global Damages

Before concluding my testimony today, I would like to address two common critiques of the social cost of carbon.

The first is that the discount rates used in the estimate are too low. Before assessing this directly, let me step back and explain why the discount rate is used.

Because CO_2 remains in the atmosphere on a timescale measured in centuries, the damages from the carbon we release today will occur over many, many decades. The discount rate allows us to translate those future damages into their present value. Put simply, using a discount rate helps us determine today's value of future environmental damages.

To provide a range of values for the social cost of carbon, the interagency working group chose to use three different discount rates—2.5 percent, 3 percent, and 5 percent per year, with the value associated with the 3 percent discount rate serving as the central value.

The use of discount rates is an appealing alternative to ad hoc decisions to only allow damages from particular years, such as only counting damages projected to occur this century. Their quantitative importance is seen when one recognizes that $100 of damages 100 years from now has a present value of $8.46, $5.20, and $0.76 when discounted at 2.5 percent, 3 percent, and 5 percent, respectively.

Of course, these are three potential discount rates, but when one opens the newspaper it is evident that there are many interest rates that could potentially be chosen. After all, the long run average nominal yield on junk bonds is 9.23 percent[8], and it is 6.17 percent on German 10-year bonds[9].

Which is the right discount rate for regulations that reduce carbon emissions? If we choose a discount rate that is too low, then we will pay too much today for mitigation efforts. If we choose a discount rate that is too high, then we will impose higher costs on our children and grandchildren than we intend.

The answer from economics is straightforward—we are best off if we use an interest rate from an investment that matches the structure of payoffs that climate mitigation provides. Thus, if the payoffs tend to appear predictably, like they do for holding a diversified portfolio of stocks, then we would want to use something like the average return for the stock market of 5.3 percent[10]. However, if the payoffs tend to appear in lean years when the economy is not growing or is even contracting, like they do for holding gold, then we would want to use a lower discount rate, likely below 3 percent.

It is worth expanding on why a low discount rate is sensible when dealing with climate damages. To give an example, consider gold. Why would anyone hold gold as an investment when its average return over the last 48 years is just 3.3 percent[11]? The answer is that investments like the stock market that pay off in relatively fat years are worth less than investments that pay off when times are tough. This is because additional income is relatively less valuable when the economy is growing. In contrast, people are willing

---

[8] 20 year average of Bank of America Merrill Lynch US High Yield Effective Yields from 1997 to 2017, retrieved from FRED, Federal Reserve Bank of St. Louis.
[9] 50 year average yield from 1965 to 2015 retrieved from FRED, Federal Reserve Bank of St. Louis.
[10] Real average annual return of S&P 500 with dividends, 1963-2012. Data from Shiller (2012): http://www.econ.yale.edu/~shiller/data.htm
[11] Real average annual return (CAGR), 1968-2016. Gold price data retrieved from FRED, Federal Reserve Bank of St. Louis.

to hold gold exactly because it is like insurance in that it does well in tough times. Additional income is more valuable when the economy isn't doing well. In other words, society's dislike of risk means that people are willing to pay a lot to protect themselves against it, and this high degree of dislike manifests itself with the very low rates of return on gold. This is a message that financial markets deliver very clearly.

A recent example comes from the Great Recession. The stock market declined by 53 percent[12], while gold increased by 14 percent[13]: gold outperformed the stock market by 67 percent. Thus, during this period of global distress, gold played the role of insurance for those investors and households who wisely hedged their exposure to major risk.

Reflecting on this example, the appropriate discount rate comes down to a judgment about whether climate change involves a substantial risk of being disruptive in a way that a significant recession or even war might be. When one considers the possibility of large temperature changes for given increases in emissions (e.g., due to higher than expected equilibrium climate sensitivity), great sea level rise in relatively short periods of time, the possibility of physical "tipping points", or human responses to these changes that include mass migration, then the case for a low discount rate appears strong. The case for using a low discount rate to determine the social cost of carbon is in many respects similar to the case for purchasing life, fire, and other insurance policies that protect against major disruptive events.

In addition to this conceptual reason to prefer low discount rates, the decline in global interest rates since the mid-1980s provide another one. The 3 percent discount rate that has been a cornerstone of regulatory analysis since 2003 draws its justification from the fact that it was roughly equal to the real rate of return on long-term government debt at that time. However, this is no longer true. For example, there has been a secular decline in the real interest rate on the 10-year Treasury note, dating back to the mid-1980s. According to a recent Council of Economic Advisors report, forecasts from the Congressional Budget Office and the Blue Chip consensus imply that the real 10-year Treasury yield is now expected to be below 2 percent.[14] The broader point is that global interest rates have declined since the social cost of carbon was set and, even setting aside the risk characteristics of payoffs from climate mitigation investments, there is a solid case that the discount rates currently used to calculate the social cost of carbon may be too high.

A second criticism of the social cost of carbon is that it measures global, rather than domestic, costs from carbon emissions. The argument goes that the task of the United States government is to improve the well-being of its citizens, and that accounting for benefits in other countries is inconsistent with that goal. The logical conclusion of this argument is that the social cost of carbon should only reflect damages that are projected to occur in the United States. However, this argument ignores the basic nature of the climate challenge, as well as the powerful political economy dynamics of the required solutions.

First, climate change is fundamentally a global, rather than domestic, phenomenon. Any country's domestic carbon emissions impose a global externality. Those emissions enter the earth's atmosphere and contribute to warming that affects the entire planet, with associated damages that vary both geographically and over time. It is undoubtedly true that challenges such as toxic spills in a U.S. river create a more straightforward calculus—the entity that imposes harm and the entity that benefits from

---

[12] S&P 500 (TR), % change from start of recession (December 2007) to lowest point (3/9/2009). (S&P Dow Jones Indices)
[13] Dow Jones Commodity Index (Gold) (TR), % change over same time period. (S&P Dow Jones Indices)
[14] Council of Economic Advisers, *Discounting for Public Policy: Theory and Recent Evidence on the Merits of Updating the Discount Rate*, 2017.

regulation are both domestic in nature. Yet, the fact that climate change does not fit neatly within this paradigm is not a justifiable cause for inaction.

This raises the second issue, which relates to the international political economy. Just as U.S. emissions contribute to global damages, each ton of CO_2 emitted outside the United States inflicts damages on the United States. Thus, we would like China, India, the European Union, and other major emitters to reduce emissions to our (and their) benefit. Yet, it is highly improbable that these countries' reductions will be just as large if we fail to account for the damages our emissions cause in their countries. This, in effect, is the classic case of a collective action problem. The point is that using global damages in calculating the social cost of carbon is likely to increase the benefits we receive in the form of greater emissions reductions abroad. In many respects, the Paris Climate Agreement, where nearly 200 countries agreed to take action on carbon emissions, is a validation of the importance of treating this as a global problem.

V. Conclusions

Ultimately, society needs to balance the costs to our economy of mitigating climate change today with climate damages. Wishing that we did not face this trade-off will not make it go away.

As the courts have underscored, the social cost of carbon provides a necessary guidepost to help us responsibly meet this balance. The best available peer-reviewed research was used to set the United States government's value of the social cost of carbon and it has since been validated by the government's own accountability office and the courts. We will be better off if a social cost of carbon based on sound science, economics, and law continues to serve as a linchpin of regulatory policy.

Michael Greenstone
Milton Friedman Professor in Economics, the College, and the Harris School, University of Chicago
Director, Energy Policy Institute at the University of Chicago (EPIC)
Director, Becker Friedman Institute for Research in Economics (as of July 1, 2017)

Michael Greenstone is the Milton Friedman Professor in Economics, the College, and the Harris School, as well as the Director of the interdisciplinary Energy Policy Institute at the University of Chicago and the Energy & Environment Lab at the University of Chicago Urban Labs. He previously served as the Chief Economist for President Obama's Council of Economic Advisers, and is a former member of the Secretary of Energy's Advisory Board. Greenstone also directed the Brookings Institution's Hamilton Project, which studies policies to promote economic growth, and has since joined its Advisory Council. He is an elected member of the American Academy of Arts and Sciences, a Fellow of the Econometric Society, and an editor of the Journal of Political Economy. Before coming to Chicago, Greenstone was the 3M Professor of Environmental Economics at MIT. He received a PhD in Economics from Princeton University.

Chairman BIGGS. Thank you, Dr. Greenstone.

I'll now recognize Dr. Michaels for five minutes to present his testimony.

TESTIMONY OF DR. PATRICK MICHAELS, PHD, DIRECTOR, CENTER FOR THE STUDY OF SCIENCE, CATO INSTITUTE; CONTRIBUTING AUTHOR TO UNITED NATIONS INTERGOVERNMENTAL PANEL ON CLIMATE CHANGE (NOBEL PEACE PRIZE 2007)

Dr. MICHAELS. May we have the first image? Thank you.

[Slide.]

Dr. MICHAELS. Mr. Chairman, Ranking Members, Members of the Subcommittees, thank you for inviting my testimony on scientific problems relating to the current calculation of the social cost of carbon or SCC. I am Patrick J. Michaels, Director of the Center for the Study of Science at the Cato Institute. Prior to that, I was a Research Professor of Environmental Sciences at University of Virginia for 30 years.

A year-and-a-half ago I testified to the Committee on Natural Resources that the Obama Administration's calculations of the SCC were in contravention of a large and growing body of scientific literature—next image——

[Slide.]

—demonstrating that the sensitivity of temperature to human emission of carbon dioxide is not nearly as large as was previously thought. And more important, the chance of a high-end warming has greatly diminished. Since then, the evidence has grown stronger.

Climate sensitivity is the amount of net warming one gets for doubling atmospheric carbon dioxide. It also roughly approximates the forecast for surface warming for the 21st century. The Obama Administration used a sensitivity specification by Roe and Baker, which is the top line there, that had a mean sensitivity of 3.0 degrees C and a 5 to 95 percent confidence limit of 1.7 to 7.14 degrees C, a very large number.

Beginning in 2011, all this work down here, a growing body of the scientific literature has yielded 32 new estimates of the sensitivity generated by more than 50 researchers from around the world with a mean sensitivity of 2 degrees C and a 5 to 95 percent confidence limit of 1.1 to 3.5 degrees C.

The large distributions of warming—next image—

[Slide.]

—used in Roe and Baker resulted in large part because of extremely wide range of estimates of the cooling effects of sulfate aerosols, another human emission. These were dramatically narrowed by researchers Nick Lewis and Judith Curry of Georgia Tech, which greatly reduced the sensitivity, as you can see here, and the spread around that sensitivity.

As my colleague Kevin Dayaratna has shown, the newer more reality-based estimates result in a dramatic lowering of the SCC. Next image.

[Slide.]

Let's now have a look at how well those climate models that were used to calculate the previous Administration's SCC are doing. This illustration is a further update of an analysis initially presented in the testimony of John Christy in 2015 with data that ended in 2014. The uptick in observed warming at the end of the record is an apparent improvement between the models and reality. But it is not. Instead, it is the 2015/2016 El Nino.

Next image.

[Slide.]

And that spread is likely to widen again in recent years, as you can see from surface temperatures here that they have dropped back down very close to their previous El Nino value.

[Slide.]

Next is a chart of predicted trends and tropical temperatures measured vertically. This is where the largest integrated warming on Earth is forecast to occur. The green line on the left is reality, which is—generally shows two to three times less warming than has been predicted. At the top of the active weather zone around here, the forecast is approximately seven times less than—or seven times more than is what is being observed. To deny this reality is to deny science.

It is the vertical temperature distribution that largely determines daily weather. If this is forecast incorrectly, then any subsidiary forecast of surface weather regimes are of little to no value. To deny that is to deny science.

There is another systematic error on the previous calculations of the SCC. We live on a planet that is becoming greener because of the direct—next image——

[Slide.]

—physiological effects of increasing carbon dioxide on plant photosynthesis. A massive survey of the scientific literature by Dr. Craig Idso shows this caused a $3.2 trillion increment in agricultural output from 1961 through 2011. My colleague Mr. Dayaratna has shown that a more realistic sensitivity in carbon dioxide fertilization can result in a negative SCC or a net external benefit from the production of carbon dioxide.

In closing, I provide you this image of the greening of our lukewarming home planet, as taken by NASA satellites. Where there are dots, the changes are statistically significant. Note that the greatest increases, the ones in pink, are in the margins of the world's deserts and the tropical rainforest, places we all feared for. To acknowledge this is to affirm reality.

Thank you very much for inviting my testimony.

[The prepared statement of Mr. Michaels follows:]

SUMMARY OF MAJOR POINTS

Patrick J. Michaels
Center for the Study of Science
Cato Institute

1. The equilibrium climate sensitivity (ECS) in the existing federal determination of the social cost of carbon is outdated and does not reflect multiple findings in recent years that the mean ECS is significantly lower, by approximately 40%, than the value used by the Obama Administration.

2. The probability distribution for the ECS in the existing federal determination of the social cost of carbon is outdated and does not reflect multiple findings in recent years that dramatically reduce the probability of an ECS of >3.5°C.

3. Satellite and balloon-sensed bulk atmospheric temperatures have warmed about half as much as was forecast since 1979 when the satellites became operational. New calculations of the social cost of carbon should take this into account.

4. Either the period 1979-present will be the most unusual period of anthropogenerated warming, or the IPCC mean ECS figure is between 50 to 33 per cent too large. New calculations of the social cost of carbon must take this into account.

5. There is an upward spike at the end of these records owing to the very strong 2015-6 El Niño that surface data show has recently dropped near to its pre- El Niño level.

6. The largest predicted warming is above the surface in the tropical atmosphere. In reality, temperatures have warmed less than half of the forecast value.

7. This error in the vertical dimension means that all model calculations of tropical rainfall changes have negative utility, and mis-specifying the vertical changes in temperature largely invalidate any forecasts of persistent changes in weather regimes.

8. The existing SCC calculations largely ignore the magnitude, or even the existence of the highly documented (and observed) enhancement of plant growth caused by increasing atmospheric carbon dioxide.

9. Satellite data confirm that the earth's surface is becoming greener, with the largest changes being on the margins of the world's great deserts. There is no accounting for this in the current calculation of the SCC.

WRITTEN STATEMENT OF

PATRICK J. MICHAELS

DIRECTOR
CENTER FOR THE STUDY OF SCIENCE
CATO INSTITUTE
WASHINGTON, DC

HEARING ON

AT WHAT COST? EXAMINING THE SOCIAL COST OF CARBON

BEFORE THE
U.S. HOUSE OF REPRESENTATIVES
COMMITTEE ON SCIENCE, SPACE, AND TECHNOLOGY
SUBCOMMITTEE ON ENVIRONMENT
SUBCOMMITTEE ON OVERSIGHT

FEBRUARY 28, 2017

Patrick J. Michaels is the director of the Center for the Study of Science at the Cato Institute. Michaels is a past president of the American Association of State Climatologists and was program chair for the Committee on Applied Climatology of the American Meteorological Society. He was a research professor of Environmental Sciences at University of Virginia for 30 years, and Virginia State Climatologist for 27 years. Michaels was a contributing author and is a reviewer of the United Nations Intergovernmental Panel on Climate Change, which was awarded the Nobel Peace Prize in 2007.

His writing has been published in the major scientific journals such as *Geophysical Research Letters, Journal of Geophysics, Climatic Change, Nature* and *Science* as well as popular serials worldwide. He is the author or editor of seven books on climate and its impact, and he was an author of the climate "paper of the year" awarded by the Association of American Geographers in 2004. He has appeared on most of the worldwide major media.

Michaels holds AB and SM degrees in biological sciences and plant ecology from the University of Chicago, and he received a PhD in ecological climatology from the University of Wisconsin at Madison in 1979.

I am Patrick J. Michaels, Director of the Center for the Study of Science at the Cato Institute, a nonprofit, non-partisan public policy research institute located here in Washington DC, and Cato is my sole source of employment income. Before I begin my testimony, I would like to make clear that my comments are solely my own and do not represent any official position of the Cato Institute.

My testimony concerns the selective science that underlies the existing federal determination of the Social Cost of Carbon and how a more inclusive and considered process would have resulted in a lower value for the social cost of carbon.

Back in 2015, the federal government's Interagency Working Group (IWG) on the Social Cost of Carbon released a report that was a response to public comments of the IWG's determination of the social cost of carbon that were solicited by the Office of Management and Budget in November 2013. Of the 140 unique sets of substantive comments received (including a set of my own), the IWG adopted none. And apart from some minor updates to its discussion on uncertainty, the IWG, in its most recent August 2016 report, retained the same, now obsolete, methodologies that were used in its initial 2010 SCC determination.

Here, I address why this decision was based on a set of flimsy, internally inconsistent excuses and amounts to a continuation of the IWG's exclusion of the most relevant science—an exclusion which assures that low, or even negative values of the social cost of carbon (which would imply a net benefit of increased atmospheric carbon dioxide levels), do not find their way into cost/benefit analyses of proposed federal actions. If, in fact, the social cost of carbon were near zero, it would eliminate the justification for any federal action (greenhouse gas emissions regulations, ethanol mandates, miles per gallon standards, solar/wind subsidies, DoE efficiency regulations, etc.) geared towards reducing carbon dioxide emissions.

Equilibrium Climate Sensitivity

In May 2013, the Interagency Working Group produced an updated SCC value by incorporating revisions to the underlying three Integrated Assessment Models (IAMs) used by the IWG in its initial 2010 SCC determination. But, at that time, the IWG did *not* update the equilibrium climate sensitivity (ECS) employed in the IAMs. This was not done, despite, now, there having been, since January 1, 2011, at least 16 new studies and 32 experiments (involving more than 50 researchers) examining the ECS, each lowering the best estimate and tightening the error distribution about that estimate. Instead, the IWG wrote in its 2013 report: "It does not revisit other interagency modeling decisions (e.g., with regard to the discount rate, reference case socioeconomic and emission scenarios, or equilibrium climate sensitivity)."

This decision was reaffirmed by the IWG in July 2015 and again in its most recent August 2016 report. But, through its reaffirmation, the IWG has again refused to give credence to and recognize the importance of what is now becoming mainstream science—that the most likely value of the equilibrium climate sensitivity is lower than that used by the IWG and that the estimate is much better constrained. This situation has profound implications for the determination of the SCC and yet continues to be summarily dismissed by the IWG.

The earth's equilibrium climate sensitivity is defined by the IWG in its 2010 report (hereafter, IWG2010) as "the long-term increase in the annual global-average surface temperature from a doubling of atmospheric CO2 concentration relative to pre-industrial levels (or stabilization at a concentration of approximately 550 parts per million (ppm))" and is recognized as "a key input parameter" for the integrated assessment models used to determine the social cost of carbon.

The IWG2010 report has an entire section (Section III.D) dedicated to describing how an estimate of the equilibrium climate sensitivity and the scientific uncertainties surrounding its actual value are developed and incorporated in the IWG's analysis. The IWG2010, in fact, developed its own probability density function (pdf) for the ECS and used it in each of the three IAMs, superseding the ECS pdfs used by the original IAMs developers. The IWG's intent was to develop an ECS pdf which most closely matched the description of the ECS as given in the *Fourth Assessment Report* of the United Nation's Intergovernmental panel on Climate Change which was published in 2007.

The functional form adopted by the IWG2010 was a calibrated version of the Roe and Baker (2007) distribution. It was described in the IWG2010 report in the following Table and Figure (from the IWG2010 report):

Table 1: Summary Statistics for Four Calibrated Climate Sensitivity Distributions

	Roe & Baker	Log-normal	Gamma	Weibull
Pr(ECS < 1.5°C)	0.013	0.050	0.070	0.102
Pr(2°C < ECS < 4.5°C)	0.667	0.667	0.667	0.667
5th percentile	1.72	1.49	1.37	1.13
10th percentile	1.91	1.74	1.65	1.48
Mode	2.34	2.52	2.65	2.90
Median (50th percentile)	3.00	3.00	3.00	3.00
Mean	3.50	3.28	3.19	3.07
90th percentile	5.86	5.14	4.93	4.69
95th percentile	7.14	5.97	5.59	5.17

67

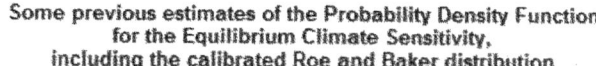

Some previous estimates of the Probability Density Function
for the Equilibrium Climate Sensitivity,
including the calibrated Roe and Baker distribution

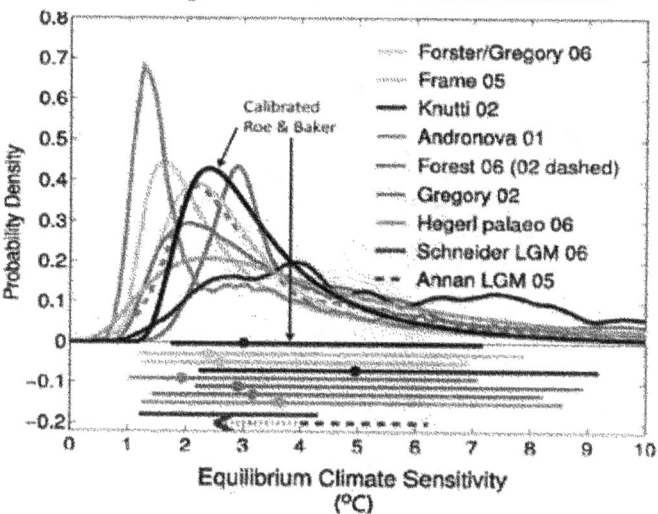

The calibrated Roe and Baker functional form used by the IWG2010 is *no longer scientifically defensible;* nor was it at the time of the publication of the IWG 2013 SCC update, nor at the time of the August 2016 update.

The figure below vividly illustrates this fact, as it compares the best estimate and 90% confidence range of the earth's ECS as used by the IWG (calibrated Roe and Baker) against findings in the scientific literature published since January 1, 2011.

Whereas the IWG ECS distribution has a median value of 3.0°C and 5th and 95th percentile values of 1.72°C and 7.14°C, respectively, the corresponding values averaged from the recent scientific literature are ~2.0°C (median), ~1.1°C (5th percentile), and ~3.5°C (95th percentile).

These differences will have large and significant impacts on the SCC determination.

68

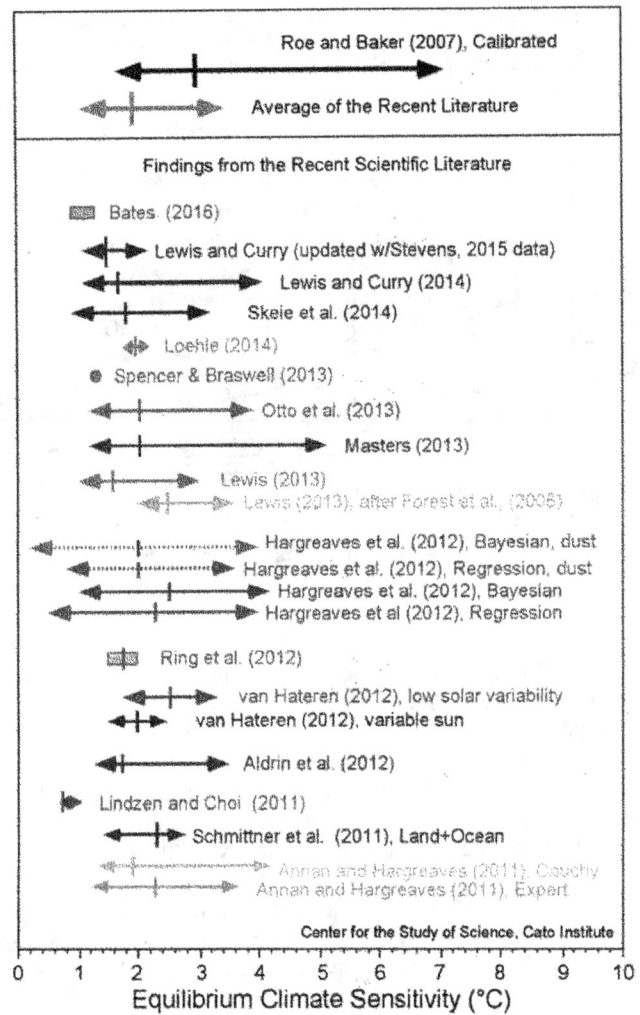

CAPTION: *The median (indicated by the small vertical line) and 90% confidence range (indicated by the horizontal line with arrowheads) of the climate sensitivity estimate used by the Interagency Working Group on the Social Cost of Carbon Climate (Roe and Baker, 2007) is indicated by the top black arrowed line. The average of the similar values from 22 different determinations reported in the recent scientific literature is given by the grey arrowed line (second line from the top). The sensitivity estimates from the 32 individual determinations of the ECS as reported in new research published after January 1, 2011 are indicated by the colored arrowed lines. The arrows indicate the 5 to 95% confidence bounds for each estimate along with the best estimate (median of each probability density function; or the mean of multiple estimates; colored vertical line). Ring et al. (2012) present four estimates of the climate sensitivity and the red box encompasses those estimates. Likewise, Bates (2016) presents eight estimates and the green box encompasses them. Spencer and Braswell (2013) produce a single ECS value best-matched to ocean heat content observations and internal radiative forcing.*

In addition to recent studies aimed at directly determining the equilibrium climate sensitivity (included in the chart above), there have been several other major studies which have produced results which qualitatively suggest a climate sensitivity lower than mainstream (e.g. Roe and Baker calibration) estimates. Such studies include new insights on cloud condensation nuclei and cosmic rays (Kirkby et al., 2016), radiative forcing of clouds (Bellouin, 2016; Stevens, 2015), cloud processes (Mauritsen and Stevens, 2015) and the underestimation of terrestrial CO_2 uptake (Sun et al., 2014).

The IWG2010 report noted that, concerning the low end of the ECS distribution, its determination reflected a greater degree of certainty that a low ECS value could be excluded than did the IPCC. From the IWG2010 (p. 14):

> "Finally, we note the IPCC judgment that the equilibrium climate sensitivity "is very likely larger than 1.5°C." Although the calibrated Roe & Baker distribution, for which the probability of equilibrium climate sensitivity being greater than 1.5°C is almost 99 percent, is not inconsistent with the IPCC definition of "very likely" as "greater than 90 percent probability," it reflects a greater degree of certainty about very low values of ECS than was expressed by the IPCC."

In other words, the IWG used its *judgment* that the lower bound of the ECS distribution was higher than the IPCC 2007 assessment indicated. However, the collection of the recent literature on the ECS shows the IWG's judgment to be in error. As can be seen in the chart above, the large majority of the findings on ECS in the recent literature indicate that the lower bound (i.e., 5th percentile) of the ECS distribution is lower than the IPCC 2007 assessment. And, the average value of the 5th percentile in the recent literature (~1.1°C) is 0.62°C *less* than that used by the IWG—a sizeable and important difference which will influence the SCC determination.

In fact, the abundance of literature supporting a lower climate sensitivity was at least partially reflected in the new IPCC assessment report issued in 2013. In that report, the IPCC reported:

> Equilibrium climate sensitivity is *likely* in the range 1.5°C to 4.5°C (*high confidence*), *extremely unlikely* less than 1°C (*high confidence*), and *very unlikely* greater than 6°C (*medium confidence*). The lower temperature limit of the assessed *likely* range is thus less than the 2°C in the AR4...

Clearly, the IWG's assessment of the low end of the probability density function that best describes the current level of scientific understanding of the climate sensitivity is incorrect and indefensible.

But even more influential in the SCC determination is the upper bound (i.e., 95th percentile) of the ECS probability distribution.

The IWG2010 notes (p.14) that the calibrated Roe and Baker distribution better reflects the IPCC judgment that "values substantially higher than 4.5°C still cannot be excluded." The IWG2010 further notes that

"Although the IPCC made no quantitative judgment, the 95th percentile of the calibrated Roe & Baker distribution (7.1 °C) is much closer to the mean and the median (7.2 °C) of the 95th percentiles of 21 previous studies summarized by Newbold and Daigneault (2009). It is also closer to the mean (7.5 °C) and median (7.9 °C) of the nine truncated distributions examined by the IPCC (Hegerl, et al., 2006) than are the 95th percentiles of the three other calibrated distributions (5.2-6.0 °C)."

In other words, the IWG2010 turned towards surveys of the scientific literature to determine its assessment of an appropriate value for the 95th percentile of the ECS distribution. Now, some seven years later, the scientific literature tells different story.

Instead of a 95th percentile value of 7.14°C, as used by the IWG2010, a survey of the recent scientific literature suggests a value of ~3.5°C—more than 50% lower.

And this is very significant and important difference because the high end of the ECS distribution has a large impact on the SCC determination—a fact frequently commented on by the IWG2010.

For example, from IWG2010 (p.26):

"As previously discussed, low probability, high impact events are incorporated into the SCC values through explicit consideration of their effects in two of the three models as well as the use of a probability density function for equilibrium climate sensitivity. Treating climate sensitivity probabilistically results in more high temperature outcomes, which in turn lead to higher projections of damages. Although FUND does not include catastrophic damages (in contrast to the other two models), its probabilistic treatment of the equilibrium climate sensitivity parameter will directly affect the non-catastrophic damages that are a function of the rate of temperature change."

And further (p.30):

Uncertainty in extrapolation of damages to high temperatures: The damage functions in these IAMs are typically calibrated by estimating damages at moderate temperature increases (e.g., DICE [Dynamic Integrated Climate and Economy] was calibrated at 2.5 °C) and extrapolated to far higher temperatures by assuming that damages increase as some power of the temperature change. Hence, estimated damages are far more uncertain under more extreme climate change scenarios.

And the entirety of Section V "A Further Discussion of Catastrophic Impacts and Damage Functions" of the IWG 2010 report describes "tipping points" and "damage functions" that are probabilities assigned to different values of global temperature change. Table 6 from the IWG2010 indicated the probabilities of various tipping points.

Table 6: Probabilities of Various Tipping Points from Expert Elicitation -

Possible Tipping Points	Duration before effect is fully realized (in years)	Additional Warming by 2100		
		0.5-1.5 C	1.5-3.0 C	3-5 C
Reorganization of Atlantic Meridional Overturning Circulation	about 100	0-18%	6-39%	18-67%
Greenland Ice Sheet collapse	at least 300	8-39%	33-73%	67-96%
West Antarctic Ice Sheet collapse	at least 300	5-41%	10-63%	33-88%
Dieback of Amazon rainforest	about 50	2-46%	14-84%	41-94%
Strengthening of El Niño-Southern Oscillation	about 100	1-13%	6-32%	19-49%
Dieback of boreal forests	about 50	13-43%	20-81%	34-91%
Shift in Indian Summer Monsoon	about 1	Not formally assessed		
Release of methane from melting permafrost	Less than 100	Not formally assessed.		

The likelihood of occurrence of these low probability, high impact, events ("tipping points") is *greatly* diminished under the new ECS findings. The average 95th percentile value of the new literature survey is only ~3.5°C indicating a very low probability of a warming reaching 3-5°C by 2100 as indicated in the 3rd column of the above Table and thus a significantly lower probability that such tipping points will be reached. This new information will have a large impact on the final SCC determination using the IWG's methodology.

The size of this impact has been directly investigated.

In their *Comment on the Landmark Legal Foundation Petition for Reconsideration of Final Rule Standards for Standby Mode and Off Mode Microwave Ovens*, Dayaratna and Kreutzer (2013) ran the DICE model using the distribution of the ECS as described by Otto et al. (2013)—a paper published in the recent scientific literature which includes 17 authors, 15 of which were lead authors of chapters in the recent Intergovernmental Panel on Climate Change's *Fifth Assessment Report*. The most likely value of the ECS reported by Otto et al. (2013) was described as "2.0°C, with a 5–95% confidence interval of 1.2–3.9°C." Using the Otto et al. (2013) ECS distribution in lieu of the distribution employed by the IWG (2013), dropped the SCC by 42 percent, 41 percent, and 35 percent (for the 2.5%, 3.0%, 5.0% discount rates, accordingly). This is a significant decline.

In subsequent research, Dayaratna and Kreutzer (2014) examined the performance of the FUND (Framework for Uncertainty, Negotiation, and Distribution) model, and found that it too, produced a greatly diminished value for the SCC when run with the Otto et al. distribution of the equilibrium climate sensitivity. Using the Otto et al. (2013) ECS distribution in lieu of the distribution employed by the IWG (2013), dropped the SCC produced by the FUND model to $11, $6, $0 compared with the original $30, $17, $2 (for the 2.5%, 3.0%, 5.0% discount rates, accordingly). Again, this is a significant decline.

The Dayaratna and Kreutzer (2014) results using FUND were in line with alternative estimates of the impact of a lower climate sensitivity on the FUND model SCC determination.

Waldhoff et al. (2011) investigated the sensitivity of the FUND model to changes in the ECS. Waldhoff et al. (2011) found that changing the ECS distribution such that the mean of the distribution was lowered from 3.0°C to 2.0°C had the effect of lowering the SCC by 60 percent (from a 2010 SCC estimate of $8/ton of CO2 to $3/ton in $1995). While Waldhoff et al. (2011) examined FUNDv3.5, the response of the current version (v3.8) of the FUND model should be similar.

Additionally, the developer of the PAGE (Policy Analysis of the Greenhouse Effect) model, affirmed that the SCC from the PAGE model, too drops by 35% when the Otto et al. (2013) climate sensitivity distribution is employed (Hope, 2013).

More recently, the FUND and DICE model were run with equilibrium climate sensitivities that were determined by Lewis and Curry (2014) in an analysis which updated and expanded upon the results of Otto et al. (2013). In Dayaratna et al. (2017), the probability density function (pdf) for the equilibrium climate sensitivity determined from an energy budget model (Lewis and Curry, 2014) was used instead of the Roe and Baker calibrated pdf used by the IWG. In doing so, Dayaranta et al. (2017) report:

> "In the DICE model the average SCC falls by 30-50% depending on the discount rate, while in the FUND model the average SCC falls by over 80%. The span of estimates across discount rates also shrinks considerably, implying less sensitivity to this parameter choice...Furthermore the probability of a negative SCC (implying CO2 emissions are a positive externality) jumps dramatically using an empirical ECS distribution."

These studies make clear that the strong dependence of the social cost of carbon on the distribution of the estimates of the equilibrium climate sensitivity (including the median, and the upper and lower certainty bounds) requires that the periodic updates to the IWG SCC determination must include a critical examination of the scientific literature on the topic of the equilibrium climate sensitivity, not merely kowtowing to the IPCC assessment. There is no indication that the IWG undertook such an independent examination. But what is clear, is that the IWG did *not* alter its probability distribution of the ECS between its 2010, 2013, 2015, and 2016 SCC determinations, despite a large and growing body of scientific literature that substantially alters and better defines the scientific understanding of the earth's ECS. It is unacceptable that a supposed "updated" social cost of carbon does not include updates to the science underlying a critical and key aspect of the SCC.

I note that there has been one prominent scientific study in the recent literature which has argued, on the basis of recent observations of lower tropospheric mixing in the tropics, for a rather high climate sensitivity (Sherwood et al., 2014). This research, however, suffers from too narrow a focus. While noting that climate models which best match the apparent observed behavior of the vertical mixing characteristics of the tropical troposphere tend to be the models with high climate sensitivity estimates, the authors fail to make note that these same models are the ones whose projections make the *worst* match to observations of the evolution of global temperature during the past several decades.

While Sherwood et al. (2014) prefer models that better match their observations in one variable, the same models actually do *worse* in the big picture than do models which lack the apparent accuracy in the processes that Sherwood et al. (2014) describe. The result can only mean that there must still be even bigger problems with *other* model processes which must more than counteract the effects of the processes described by Sherwood et al.

This illustrates the inherent problems with "tuning" climate models to try to best reproduce a known set of observations—efforts to force climate models to better emulate one set of physical behaviors can degrade their performance on other ones. Voosen (2016) recently reported on the climate modelling community efforts to be more open and transparent with their multitude of (secret) "tuning" procedures. Voosen's reporting was eye-opening not only in revealing the degree to which climate models are tuned and the significant role that tuning plays in model projections, but as to the reasons why modelers have not wanted to be up front about their methods. I reproduce an extended and relevant excerpt here:

> At their core, climate models are about energy balance. They divide Earth up into boxes, and then, applying fundamental laws of physics, follow the sun's energy as it drives phenomena like winds and ocean currents. Their resolution has grown over the years, allowing current models to render Earth in boxes down to 25 kilometers a side. They take weeks of supercomputer time for a full run, simulating how the climate evolves over centuries.

> When the models can't physically resolve certain processes, the parameters take over—though they are still informed by observations. For example, modelers tune for cloud formation based on temperature, atmospheric stability, humidity, and the presence of mountains. Parameters are also used to describe the spread of heat into the deep ocean, the reflectivity of Arctic sea ice, and the way that aerosols, small particles in the atmosphere, reflect or trap sunlight.

> It's impossible to get parameters right on the first try. And so scientists adjust these equations to make sure certain constraints are met, like the total energy entering and leaving the planet, the path of the jet stream, or the formation of low marine clouds off the California coast. Modelers try to restrict their tuning to as few knobs as possible, but it's never as few as they'd like. It's an art and a science. "It's like reshaping an instrument to compensate for bad sound," Stevens says.

> Indeed, whether climate scientists like to admit it or not, nearly every model has been calibrated precisely to the 20th century climate records—otherwise it would

have ended up in the trash. "It's fair to say all models have tuned it," says Isaac Held, a scientist at the Geophysical Fluid Dynamics Laboratory, another prominent modeling center, in Princeton, New Jersey.

For years, climate scientists had been mum in public about their "secret sauce": What happened in the models stayed in the models. The taboo reflected fears that climate contrarians would use the practice of tuning to seed doubt about models— and, by extension, the reality of human-driven warming. "The community became defensive," [Bjorn] Stevens [of the Max Planck Institut] says. "It was afraid of talking about things that they thought could be unfairly used against them." Proprietary concerns also get in the way. For example, the United Kingdom's Met Office sells weather forecasts driven by its climate model. Disclosing too much about its code could encourage copycats and jeopardize its business.

But modelers have come to realize that disclosure could reveal that some tunings are more deft or realistic than others. It's also vital for scientists who use the models in specific ways. They want to know whether the model output they value—say, its predictions of Arctic sea ice decline— arises organically or is a consequence of tuning. [Gavin] Schmidt [Head of NASA's Goddard Institute for Space Studies, which, ironically concentrates on earth's climate] points out that these models guide regulations like the U.S. Clean Power Plan, and inform U.N. temperature projections and calculations of the social cost of carbon. "This isn't a technical detail that doesn't have consequence," he says. "It has consequence."

Recently, while preparing for the new model comparisons, MPIM modelers got another chance to demonstrate their commitment to transparency. They knew that the latest version of their model had bugs that meant too much energy was leaking into space. After a year spent plugging holes and fixing it, the modelers ran a test and discovered something disturbing: The model was now overheating. Its climate sensitivity—the amount the world will warm under an immediate doubling of carbon dioxide concentrations from preindustrial levels—had shot up from 3.5°C in the old version to 7°C, an implausibly high jump.

MPIM hadn't tuned for sensitivity before— it was a point of pride—but they had to get that number down. Thorsten Mauritsen, who helps lead their tuning work, says he tried tinkering with the parameter that controlled how fast fresh air mixes into clouds. Increasing it began to ratchet the sensitivity back down. "The model we produced with 7° was a damn good model," Mauritsen says. But it was not the team's best representation of the climate as they knew it.

That climate modelers were worried about being open about their methodologies for fear that "contrarians" would "unfairly" use such procedures against them indicates that the modeling community is more interested in climate policy (that may find support in their model projections) than climate science (which would welcome criticism aimed at producing a better understanding of the physical processes driving the earth's climate). Given the degree of "secret sauce" mixed into the models at this point in time, a healthy dose of skepticsm regarding the verisimilitude of climate model output is warranted.

But even with the all the model tuning that takes place, the overall model collective is *still* warming the world much faster than it actually is. As shown by Christy (2016, and updates), there is a gross departure of "reality" from model predictions. Christy (2016) noted that "for the global bulk troposphere [roughly the bottom 40,000 feet of the atmosphere], the models overwarm the atmosphere by a factor of about 2.5." The warming influence of a large and naturally occurring El Niño event has, temporarily, added a blip to the end of the observational record. But despite this short-term natural warming event, collectively the models still produce about twice as much warming as can be found in the real world over the past 38 years. And as the warming of the recent strong El Niño event fades (global surface temperatures have returned most of the way to pre-El Niño levels; see Figures below), the model/real world discrepancy will start to grow once again.

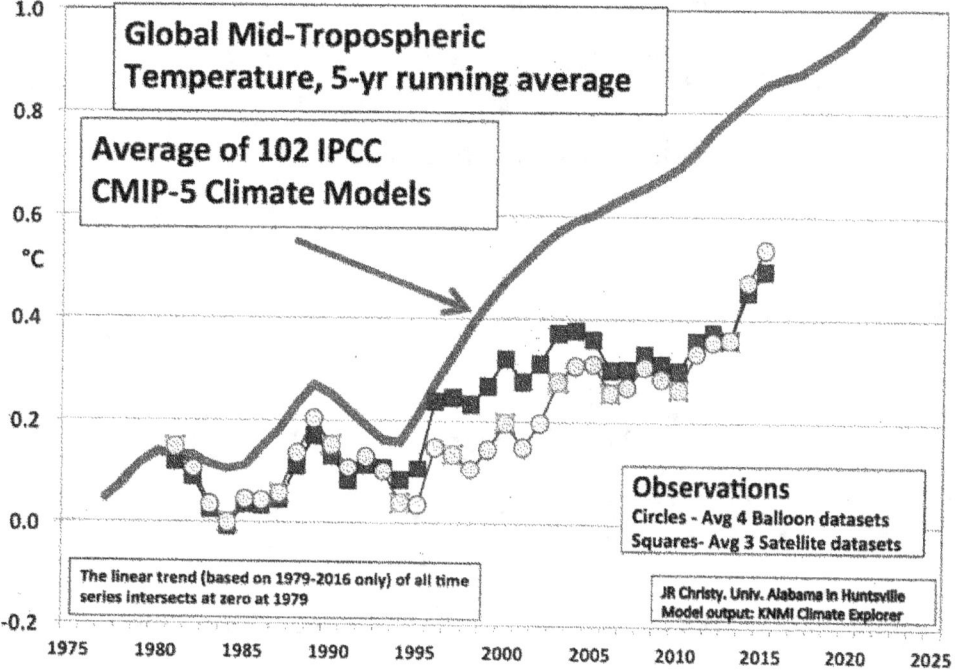

CAPTION: *Five-year running mean temperatures predicted by the UN's climate models, and observed lower atmospheric temperatures from weather balloons and satellites (figure courtesy of John Christy). The last point is a four-year running mean, and the first two are three and four, respectively.*

Surface Observations (HadCRUT4v5)

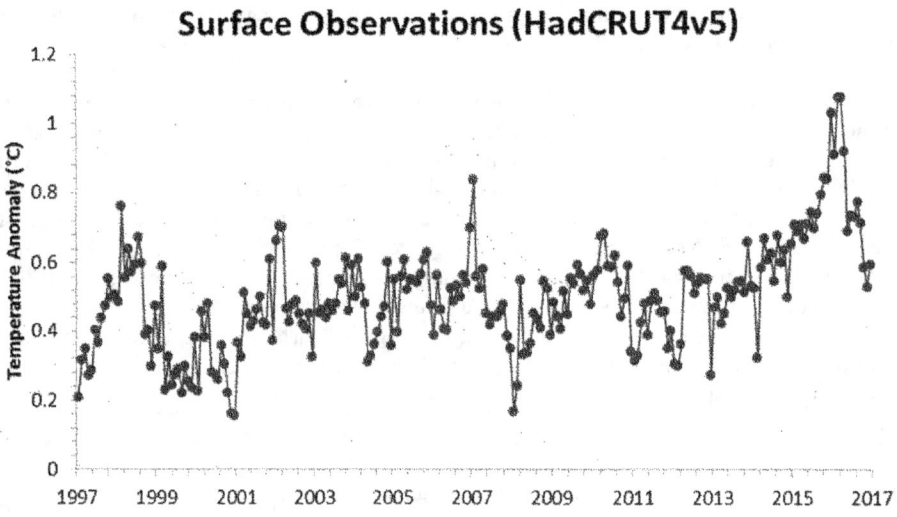

Satellite Observations (UAH_MT)

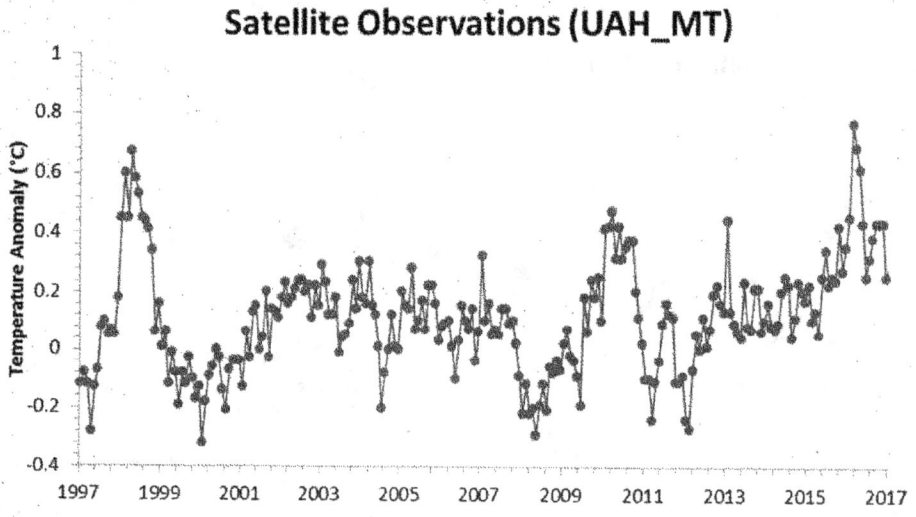

CAPTION: Monthly temperature anomalies, January 1997 through December 2016 (surface observations; top) and January 1997 through January 2017 (satellite observations of the mid-troposphere; bottom) show the impact of the strong 2016 El Niño event and the fading warmth since. The surface readings are from the Climate Research Unit at the University of East Anglia, and the satellite readings are from University of Alabama-Huntsville.

Another way to assess model performance is to compare model projection with observed trends in the vertical dimension of the atmosphere. Here again, as shown in the Figure below, models grossly produce much more warming than has been observed. This chart, courtesy of the University of Alabama at Huntsville's Dr. John Christy, focuses on the tropics (between 20S and 20N)—the area where climate models project the greatest amount of warming through the atmosphere. The communal failure of the models is abject.

The characteristics of the vertical profile of temperature are important environmental variables in that it is the vertical temperature distribution that determines atmospheric stability. When the lapse rate—the difference between the lowest layers and higher levels—is large, the atmosphere is unstable. Instability is the principal source for global precipitation. Although models can be (and are) tuned to mimic changes in surface temperatures, the same can't be done as easily for the vertical profile of temperature changes. As the figure indicates, the air in the middle troposphere is warming far more slowly than has been predicted, even more slowly than the torpid surface warming. Consequently, the difference between the surface and the middle troposphere has become slightly greater, a condition which should produce a very slight increase in average precipitation. On the other hand, the models forecast that the difference between the surface and the middle troposphere should become less, a condition which would add pressure to decrease global precipitation.

The models are therefore making systematic errors in their precipitation projections. That has a dramatic effect on the resultant climate change projections. When the surface is wet, which is what occurs after rain, the sun's energy is directed toward the evaporation of that moisture rather than to directly heating the surface. In other words, much of what is called "sensible weather" (the kind of weather a person can sense) is determined by the vertical distribution of temperature. If the popular climate models get that wrong (which is what is happening), then all the subsidiary weather may also be incorrectly specified.

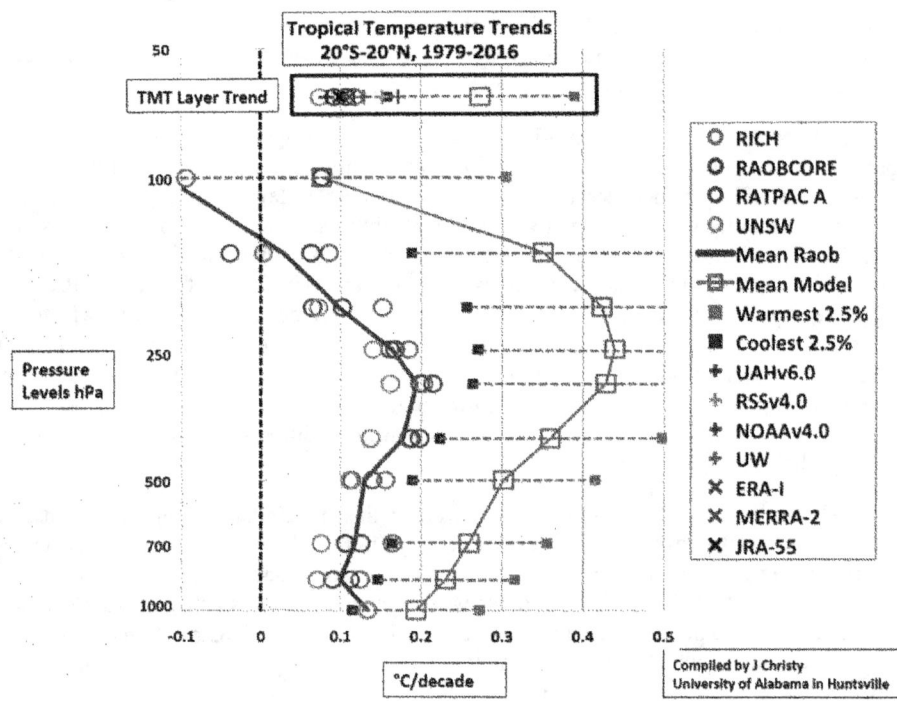

CAPTION: Tropical (20°S to 20°N) temperature trends(1979-2016) throughout the vertical atmosphere as projected by climate models (red squares, with uncertainty) and as observed by radiosondes carried aloft by weather balloons (colored circles represent different data compilations). The red line is the model mean and the green line is the observed mean. The trend in the bulk lower atmosphere (middle troposphere) from several different satellite data compilations (colored plus signs, top box) and several reanalysis datasets (colored crosses, top box) is compared with the model projection for the same layer in the box at the top of the figure. (Figure courtesy of John Christy)

These results argue strongly against the reliability of the Sherwood et al. (2014) conclusion and instead provide robust observational evidence that the climate sensitivity has been overestimated by both climate models, and the IWG alike.

Agricultural Impacts of Carbon Fertilization

Carbon dioxide is known to have a large positive impact on vegetation (e.g., Zhu et al., 2016), with literally thousands of studies in the scientific literature demonstrating that plants (including crops) grow stronger, healthier, and more productive under conditions of increased carbon dioxide concentration. A study (Idso, 2013) reviewed a large collection of such literature as it applies to the world's 45 most important food crops (making up 95% of the world's annual agricultural production).

Idso (2013) summarized his findings on the increase in biomass of each crop that results from a 300ppm increase in the concentration of carbon dioxide under which the plants were grown. This table is reproduced below, and shows that the typical growth increase exceeds 30% in most crops, including 8 of the world's top 10 food crops (the increase was 24% and 14% in the other two).

Idso (2013) found that the increase in the atmospheric concentration of carbon dioxide that took place during the period 1961-2011 was responsible for increasing global agricultural output by 3.2 trillion dollars (in 2004-2006 constant dollars). Projecting the increases forward based on projections of the increase in atmospheric carbon dioxide concentration, Idso (2013) expects carbon dioxide fertilization to increase the value of agricultural output by 9.8 trillion dollars (in 2004-2006 constant dollars) during the 2012-2050 period.

Average percentage increase in biomass of each of the world's 45 most important food crops under an increase of 300ppm of carbon dioxide.

Crop	% Biomass Change	Crop	% Biomass Change
Sugar cane	34.0%	Rye	38.0%
Wheat	34.9%	Plantains	44.8%
Maize	24.1%	Yams	47.0%
Rice, paddy	36.1%	Groundnuts, with shell	47.0%
Potatoes	31.3%	Rapeseed	46.9%
Sugar beet	65.7%	Cucumbers and gherkins	44.8%
Cassava	13.8%	Mangoes, mangosteens, guavas	36.0%
Barley	35.4%	Sunflower seed	36.5%
Vegetables fresh nes	41.1%	Eggplants (aubergines)	41.0%
Sweet potatoes	33.7%	Beans, dry	61.7%
Soybeans	45.5%	Fruit Fresh Nes	72.3%
Tomatoes	35.9%	Carrots and turnips	77.8%
Grapes	68.2%	Other melons (inc.cantaloupes)	4.7%
Sorghum	19.9%	Chillies and peppers, green	41.1%
Bananas	44.8%	Tangerines, mandarins, clem.	29.5%
Watermelons	41.5%	Lettuce and chicory	18.5%
Oranges	54.9%	Pumpkins, squash and gourds	41.5%
Cabbages and other brassicas	39.3%	Pears	44.8%
Apples	44.8%	Olives	35.2%
Coconuts	44.8%	Pineapples	5.0%
Oats	34.8%	Fruit, tropical fresh nes	72.3%
Onions, dry	20.0%	Peas, dry	29.2%
Millet	44.3%		

This is a large positive externality, and one that is insufficiently modeled in the IAMs relied upon by the IWG in determining the SCC.

In fact, only one of the three IAMs used by the IWG has any substantial impact from carbon dioxide fertilization, and the one that does, underestimates the effect by approximately 2-3 times.

The FUND model has a component which calculates the impact on agricultural as a result of carbon dioxide emissions, which includes not only the impact on temperature and other climate changes, but also the direct impact of carbon dioxide fertilization. The other two IAMs, DICE and PAGE by and large do not (or only do so extremely minimally; DICE includes the effect to a larger degree than PAGE). Consequently, lacking this large and positive externality, the SCC calculated by the DICE and PAGE models is significantly larger than the SCC determined by the FUND model (for example, see Table A5, in the IWG 2013 report).

But even the positive externality that results from carbon dioxide fertilization as included in the FUND model is too small when compared with the Idso (2013) estimates. FUND (v3.7) uses the following formula to determine the degree of crop production increase resulting from atmospheric carbon dioxide increases (taken from Anthoff and Tol, 2013a):

CO$_2$ fertilisation has a positive, but saturating effect on agriculture, specified by

(A.4)
$$A_{t,r}^f = \gamma_r \ln \frac{CO2_t}{275}$$

where

- A^f denotes damage in agricultural production as a fraction due to the CO2 fertilisation by time and region;

- t denotes time;

- r denotes region;

- $CO2$ denotes the atmospheric concentration of carbon dioxide (in parts per million by volume);

- 275 ppm is the pre-industrial concentration;

- γ is a parameter (see Table A, column 8-9).

Column 8 in the table below shows the CO$_2$ fertilization parameter (γ_r) used in FUND for various regions of the world (Anthoff and Tol, 2013b). The average CO$_2$ fertilization effect across the 16 regions of the world is 11.2%. While this number is neither areally weighted, nor weighted by the specific crops grown, it is clear that 11.2% is much lower than the average fertilization effect compiled by Idso (2013) for the world's top 10 food crops (35%). Further, Idso's fertilization impact is in response to a 300ppm CO2 increase, while the fertilization parameter in the FUND model is multiplied by ln(CO2$_t$/275) which works out to 0.74 for a 300ppm CO$_2$ increase. This multiplier further reduces the 16 region average to 8.4% for the CO$_2$ fertilization effect—some 4 times smaller than the magnitude of the fertilization impact identified by Idso (2013).

Although approximately four times too small, the impact of the fertilization effect on the SCC calculation in the FUND model is large.

According to Waldhoff et al. (2011), if the CO$_2$ fertilization effect is turned off in the FUND model (v3.5) the SCC increases by 75% from \$8/tonCO$_2$ to \$14/tonCO$_2$ (in 1995 dollars). In another study, Ackerman and Munitz (2012) find the effective increase in the FUND model to be even larger, with CO$_2$ fertilization producing a positive externality of nearly \$15/tonCO$_2$ (in 2007 dollars).

Impact of climate change on agriculture in FUND model.

	Rate of change (% Ag. Prod/ 0.04°C)		δ_r^l		δ_r^q		CO_2 fertilisation (% Ag. Prod)	
USA	-0.021	(0.176)	0.026	(0.021)	-0.012	(0.018)	8.90	(14.84)
CAN	-0.029	(0.073)	0.092	(0.080)	-0.016	(0.009)	4.02	(6.50)
WEU	-0.039	(0.138)	0.022	(0.002)	-0.014	(0.013)	15.41	(11.83)
JPK	-0.033	(0.432)	0.046	(0.022)	-0.024	(0.030)	23.19	(36.60)
ANZ	-0.015	(0.142)	0.040	(0.071)	-0.016	(0.037)	10.48	(8.50)
EEU	-0.027	(0.062)	0.048	(0.097)	-0.018	(0.048)	9.52	(5.14)
FSU	-0.018	(0.066)	0.042	(0.075)	-0.016	(0.039)	6.71	(5.48)
MDE	-0.022	(0.032)	0.042	(0.071)	-0.017	(0.037)	9.43	(2.66)
CAM	-0.034	(0.061)	0.064	(0.043)	-0.030	(0.043)	16.41	(5.38)
SAM	-0.009	(0.060)	0.003	(0.005)	-0.004	(0.003)	5.96	(5.04)
SAS	-0.014	(0.021)	0.025	(0.024)	-0.011	(0.018)	5.80	(1.64)
SEA	-0.009	(0.482)	0.014	(0.004)	-0.010	(0.008)	8.45	(41.81)
CHI	-0.013	(0.075)	0.043	(0.076)	-0.017	(0.040)	19.21	(6.13)
NAF	-0.016	(0.023)	0.033	(0.043)	-0.014	(0.027)	7.27	(1.90)
SSA	-0.011	(0.026)	0.024	(0.034)	-0.010	(0.020)	5.05	(2.20)
SIS	-0.050	(0.103)	0.043	(0.077)	-0.017	(0.040)	23.77	(8.64)

Standard deviations are given in brackets.

Clearly, had the Idso (2013) estimate of the CO_2 fertilization impact been used instead of the one used in FUND the resulting positive externality would have been much larger, and the resulting net SCC been much lower.

This is just for one of the three IAMs used by the IWG. Had the more comprehensive CO_2 fertilization impacts identified by Idso (2013) been incorporated in all the IAMs, the three-model average SCC used by the IWG would be been greatly lowered, and likely even become negative in some IAM/discount rate combinations.

In its 2015 "Response to Comments Social Cost of Carbon for Regulatory Impact Analysis Under Executive Order 12866," the IWG admits to the disparate ways that CO_2 fertilization is included in the three IAMs. Nevertheless, the IWG quickly dismisses this as a problem in that they claim the IAMs were selected "to reflect a reasonable range of modeling choices and approaches that collectively reflect the current literature on the estimation of damages from CO2 emissions."

This logic is blatantly flawed. Two of the IAMs do not reflect the "current literature" on a key aspect relating to the direct impact of CO2 emissions on agricultural output, and the third only partially so.

CO_2 fertilization is a known physical effect from increased carbon dioxide concentrations. By including the results of IAMs that do not include known processes that have a significant impact on the end product must disqualify them from contributing to the final result. The inclusion of results that are known *a priori* to be wrong can only contribute to producing a less accurate

answer. Results should only be included when they attempt to represent known processes, not when they leave those processes out entirely.

The justification from the IWG (2015) that "[h]owever, with high confidence the IPCC (2013) stated in its Fifth Assessment Report (AR5) that '[b]ased on many studies covering a wide range of regions and crops, negative impacts of climate change on crop yields have been more common than positive ones'" is completely irrelevant as CO_2 fertilization is an impact that is apart from "climate change." And further, the IAMs do (explicitly in the case of FUND and DICE or implicitly in the case of PAGE) include damage functions related to the climate change impacts on agriculture. So not only is the IWG justification irrelevant, it is inaccurate as well. The impact of CO_2 fertilization on agricultural output and its impact on lowering the SCC *must* be considered.

Additional Climate Model Parameter Misspecifications

In addition to the outdated climate sensitivity distribution and the insufficient handling of the carbon dioxide fertilization effect, there has also been identified a misspecification of some of the critical parameters within the underlying box models that drive the pace and shape of the future climate evolution in the IAMs.

A recent analysis (Lewis, 2016) finds that the physically-based two-box climate model inherent in the DICE IAM is fit with physically unrealistic ocean characteristics. According to Lewis (2016):

> In the DICE 2-box model, the ocean surface layer that is taken to be continuously in equilibrium with the atmosphere is 550 m deep, compared to estimates in the range 50–150 m based on observations and on fitting 2-box models to AOGCM responses. The DICE 2-box model's deep ocean layer is less than 200 m deep, a fraction of the value in any CMIP5 AOGCM, and is much more weakly coupled to the surface layer. Unsurprisingly, such parameter choices produce a temperature response time profile that differs substantially from those in AOGCMs and in 2-box models with typical parameter values. As a result, DICE significantly overestimates temperatures from the mid-21st century on, and hence overestimates the SCC and optimum carbon tax, compared with 2-box models having the same ECS and TCR but parameter values that produce an AOGCM-like temperature evolution.

When the DICE 2-box model is parametrized with values for the ocean layers that are in line with established estimates, the value of the social cost of carbon that results is reduced by one-quarter to one-third during the 21th century. Lewis further point out that notes that "The climate response profile in FUND and in PAGE, the other two IAMs used by the US government to assess the SCC, appear to be similarly inappropriate, suggesting that they also overestimate the SCC."

Ultimately, Lewis (2016) concludes:

It seems rather surprising that all three of the main IAMs have climate response functions with inappropriate, physically unrealistic, time profiles. In any event, it is worrying that governments and their scientific and economic advisers have used these IAMs and, despite considering what [equilibrium climate sensitivity] and/or [transient climate sensitivity] values or probability distributions thereof to use, have apparently not checked whether the time profiles of the resulting climate responses were reasonable.

Sea Level Rise

The sea level rise module in the DICE model used by the IWG2013/2015/2016 produces future sea level rise values that far exceed mainstream projections and are unsupported by the best available science. The sea level rise projections from more than half of the scenarios (IMAGE, MERGE, MiniCAM) exceed even the highest end of the projected sea level rise by the year 2300 as reported in the *Fifth Assessment Report* (AR5) of the Intergovernmental Panel on Climate Change (see figure).

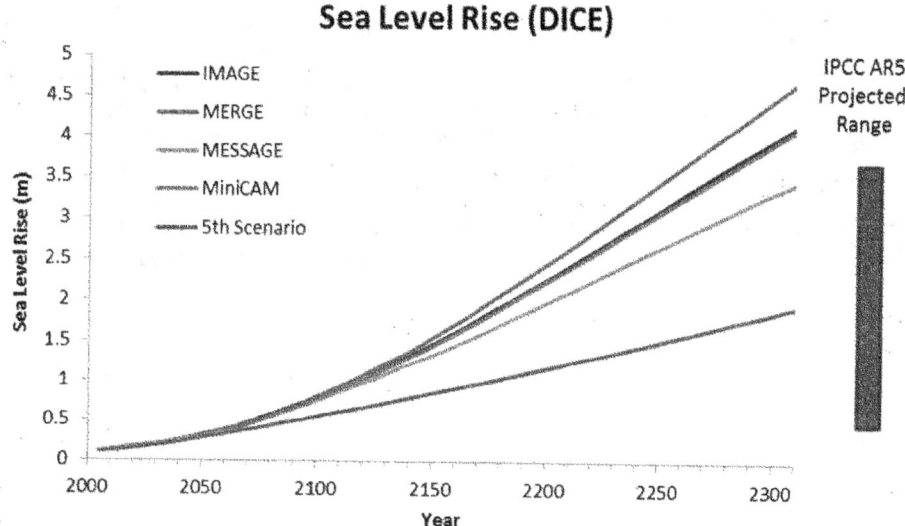

CAPTION: *Projections of sea level rise from the DICE model (the arithmetic average of the 10,000 Monte Carlo runs from each scenario) for the five scenarios examined by the IWG2013 compared with the range of sea level rise projections for the year 2300 given in the IPCC AR5 (see AR5 Table 13.8). (DICE data provided by Kevin Dayaratna and David Kreutzer of the Heritage Foundation).*

How the sea level rise module in DICE was constructed is inaccurately characterized by the IWG (and misleads the reader). The IWG report describes the development of the DICE sea level rise scenario as:

"The parameters of the four components of the SLR module are calibrated to match consensus results from the IPCC's Fourth Assessment Report (AR4).[6]"

However, in IWG footnote "6" the methodology is described this way (Nordhaus, 2010):

"The methodology of the modeling is to use the estimates in the IPCC Fourth Assessment Report (AR4)."

"Using estimates" and "calibrating" are two completely different things. Calibration implies that the sea level rise estimates produced by the DICE sea level module behave similarly to the IPCC sea level rise projections and instills a sense of confidence in the casual reader that the DICE projections are in accordance with IPCC projections. However this is not the case. Consequently, the reader is misled.

In fact, the DICE estimates are much higher than the IPCC estimates. This is even recognized by the DICE developers. From the same reference as above:

"The RICE [DICE] model projection is in the middle of the pack of alternative specifications of the different Rahmstorf specifications. Table 1 shows the RICE, base Rahmstorf, and average Rahmstorf. *Note that in all cases, these are significantly above the IPCC projections in AR4.*" [emphasis added]

That the DICE sea level rise projections are far above mainstream estimated can be further evidenced by comparing them with the results produced by the IWG-accepted MAGICC modelling tool (in part developed by the EPA and available from http://www.cgd.ucar.edu/cas/wigley/magicc/).

Using the MESSAGE scenario as an example, the sea level rise estimate produced by MAGICC for the year 2300 is 1.28 meters—a value that is less than 40% of the average value of 3.32 meters produced by the DICE model when running the same scenario (see figure below).

Projected Sea Level Rise (MESSAGE)

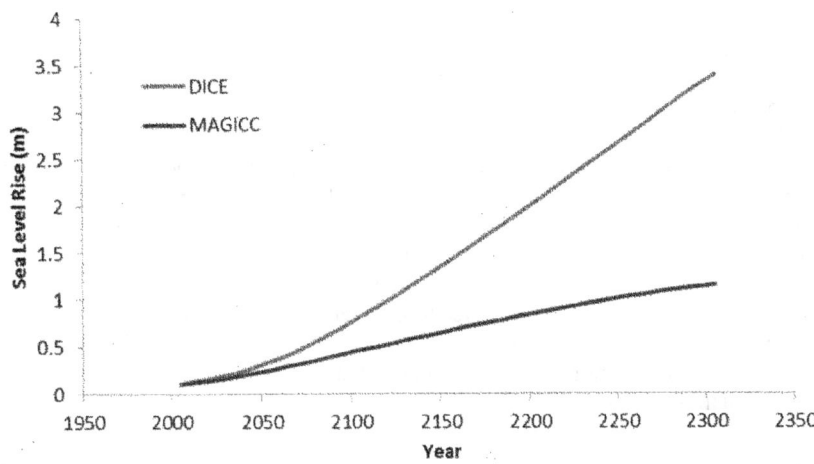

CAPTION: Projected sea level rise resulting from the MESSAGE scenario produced by DICE (red) and MAGICC (blue).

The justification given for the high sea level rise projections in the DICE model (Nordhaus, 2010) is that they well-match the results of a "semi-empirical" methodology employed by Rahmstorf (2007) and Vermeer and Rahmstorf (2009).

However, subsequent science has proven the "semi-empirical" approach to projecting future sea level rise unreliable. For example, Gregory et al. (2012) examined the assumption used in the "semi-empirical" methods and found them to be unsubstantiated. Gregory et al (2012) specifically refer to the results of Rahmstorf (2007) and Vermeer and Rahmstorf (2009):

> The implication of our closure of the [global mean sea level rise, GMSLR] budget is that a relationship between global climate change and the rate of GMSLR is weak or absent in the past. The lack of a strong relationship is consistent with the evidence from the tide-gauge datasets, whose authors find acceleration of GMSLR during the 20th century to be either insignificant or small. It also calls into question the basis of the semi-empirical methods for projecting GMSLR, which depend on calibrating a relationship between global climate change or radiative forcing and the rate of GMSLR from observational data (Rahmstorf, 2007; Vermeer and Rahmstorf, 2009; Jevrejeva et al., 2010).

In light of these findings, the justification for the very high sea level rise projections (generally exceeding those of the IPCC AR5 and far greater than the IWG-accepted MAGICC results) produced by the DICE model is called into question and can no longer be substantiated.

Given the strong relationship between sea level rise and future damage built into the DICE model, there can be no doubt that the SCC estimates from the DICE model are higher than the best science would allow and consequently, should not be accepted by the IWG as a reliable estimate of the social cost of carbon.

And here again, the IWG (2015) admits that these sea level rise estimates are an outlier on the high end, yet retains them in their analysis by claiming than they were interested in representing a "range" of possible outcomes. But, even the IWG (2015) admits that the IPCC AR5 assigned "a low confidence in projections based on such [semi-empirical] methods." It is internally inconsistent to claim the IPCC as an authority for limiting the range of possibilities explored by the IAMs (which it did in the case of equilibrium climate sensitivity) and then go outside the IPCC to justify including a wildly high estimate of sea level rise. Such inconsistencies characterize the IWG response to comments and weaken confidence in them.

I did not investigate the sea level rise projections from the FUND or the PAGE model, but suggest that such an analysis must be carried out prior to extending any confidence in the values of the SCC resulting from those models—confidence that, as demonstrated, cannot be assigned to the DICE SCC determinations.

Conclusion

The social cost of carbon as determined by the Interagency Working Group in their August 2016 Technical Support Document (updated from IGW reports from February 2010, November 2013, and July 2015) is unsupported by the robust scientific literature, fraught with uncertainty, illogical, and thus completely unsuitable and inappropriate for federal rulemaking. Had the IWG included a better-reasoned and more inclusive review of the current scientific literature, the social cost of carbon estimates would have been considerably reduced with a value likely approaching zero. Such a low social cost of carbon would obviate the arguments behind the push for federal greenhouse gas regulations.

References

Ackerman, F., and C. Munitz, 2012. Climate damages in the FUND model: a disaggregated analysis. *Ecological Economics*, **77**, 219-224.

Aldrin, M., et al., 2012. Bayesian estimation of climate sensitivity based on a simple climate model fitted to observations of hemispheric temperature and global ocean heat content. *Environmetrics*, doi: 10.1002/env.2140.

Annan, J.D., and J.C Hargreaves, 2011. On the generation and interpretation of probabilistic estimates of climate sensitivity. *Climatic Change*, **104**, 324-436.

Anthoff, D., and R.S.J. Tol, 2013a. The climate framework for uncertainty, negotiation and distribution (FUND), technical description, version 3.7, http://www.fund-model.org/publications

Anthoff, D., and R.S.J. Tol, 2013b. The climate framework for uncertainty, negotiation and distribution (FUND), tables, version 3.7, http://www.fund-model.org/publications

Bates, J. R., 2016. Estimating Climate Sensitivity Using Two-zone Energy Balance Models. *Earth and Space Science*, doi: 10.1002/2015EA000154

Bellouin, N., 2016. The interaction between aerosols and clouds. Weather and Climate @Reading, http://blogs.reading.ac.uk/weather-and-climate-at-reading/2016/1053/.

Christy, J.R., 2016. Testimony before the House Committee on Science, Space, and Technology, February 2, 2016.

Dayaratna, K., and D. Kreutzer, 2013. Comment on the Energy Efficiency and Renewable Energy Office (EERE) Proposed Rule: 2013-08-16 Energy Conservation Program for Consumer Products: Landmark Legal Foundation; Petition for Reconsideration; Petition for Reconsideration; Request for Comments, http://www.regulations.gov/#!documentDetail;D=EERE-2013-BT-PET-0043-0024

Dayaratna, K., and D. Kreutzer, 2014. Unfounded FUND: Yet another EPA model not ready for the Big Game, http://www.heritage.org/research/reports/2014/04/unfounded-fund-yet-another-epa-model-not-ready-for-the-big-game.

Dayaratna, K., R. McKitrick, and D. Kreutzer, 2017. Empirically-constrained climate sensitivity and the social cost of carbon. Accepted, *Climate Change Economics*.

Gregory, J., et al., 2012. Twentieth-century global-mean sea-level rise: is the whole greater than the sum of the parts? *Journal of Climate*, doi:10.1175/JCLI-D-12-00319.1, in press.

Hargreaves, J.C., et al., 2012. Can the Last Glacial Maximum constrain climate sensitivity? *Geophysical Research Letters*, **39**, L24702, doi: 10.1029/2012GL053872

Hope, C., 2013. How do the new estimates of transient climate response affect the social cost of CO2? http://www.chrishopepolicy.com/2013/05/how-do-the-new-estimates-of-transient-climate-response-affect-the-social-cost-of-co2/ (last visited May 5, 2014)

Idso, C. 2013. *The positive externalities of carbon dioxide: Estimating the monetary benefits of rising CO2 concentrations on global food production.* Center for the Study of Carbon Dioxide and Global Change, 30pp.

Intergovernmental Panel on Climate Change, 2007. *Climate Change 2007: The Physical Science Basis. Contribution of Working Group I to the Fourth Assessment Report of the Intergovernmental Panel on Climate Change.* Solomon, S., et al. (eds). Cambridge University Press, Cambridge, 996pp.

Intergovernmental Panel on Climate Change, 2013. *Climate Change 20013: The Physical Science Basis. Contribution of Working Group I to the Fifth Assessment Report of the Intergovernmental Panel on Climate Change.* Final Draft Accepted in the 12th Session of

Working Group I and the 36th Session of the IPCC on 26 September 2013 in Stockholm, Sweden.

Kirkby, J., et al., 2016. Ion-induced nucleation of pure biogenic particles. *Nature,* 533, 521–526, doi:10.1038/nature17953.

Lewis, N. 2013. An objective Bayesian, improved approach for applying optimal fingerprint techniques to estimate climate sensitivity. *Journal of Climate*, doi: 10.1175/JCLI-D-12-00473.1.

Lewis, N. and J.A. Curry, C., 2014. The implications for climate sensitivity of AR5 forcing and heat uptake estimates. *Climate Dynamic*, 10.1007/s00382-014-2342-y.

Lewis, N., 2016. Abnormal climate response of the DICE IAM – a trillion dollar error? *Climate Etc.*, last accessed, August 16, 2016, https://judithcurry.com/2016/08/15/abnormal-climate-response-of-the-dice-iam-a-trillion-dollar-error/

Lindzen, R.S., and Y-S. Choi, 2011. On the observational determination of climate sensitivity and its implications. *Asia-Pacific Journal of Atmospheric Science,* 47, 377-390.

Loehle, C., 2014. A minimal model for estimating climate sensitivity. *Ecological Modelling,* 276, 80-84.

Masters, T., 2013. Observational estimates of climate sensitivity from changes in the rate of ocean heat uptake and comparison to CMIP5 models. *Climate Dynamics*, doi:101007/s00382-013-1770-4

Mauritsen, T. and B. Stevens, 2015. Missing iris effect as a possible cause of muted hydrological change and high climate sensitivity in models. *Nature Geoscience*, 8, 346-351, doi:10.1038/ngeo2414.

Nordhaus, W., 2010. Projections of Sea Level Rise (SLR), http://www.econ.yale.edu/~nordhaus/homepage/documents/SLR_021910.pdf

Otto, A., F. E. L. Otto, O. Boucher, J. Church, G. Hegerl, P. M. Forster, N. P. Gillett, J. Gregory, G. C. Johnson, R. Knutti, N. Lewis, U. Lohmann, J. Marotzke, G. Myhre, D. Shindell, B. Stevens, and M. R. Allen, 2013. Energy budget constraints on climate response. *Nature Geoscience*, 6, 415-416.

Rahmstorf, S., 2007. A semi-empirical approach to projecting future sea-level rise. *Science*, 315, 368–370, doi:10.1126/science.1135456.

Ring, M.J., et al., 2012. Causes of the global warming observed since the 19th century. *Atmospheric and Climate Sciences,* 2, 401-415, doi: 10.4236/acs.2012.24035.

Schmittner, A., et al. 2011. Climate sensitivity estimated from temperature reconstructions of the Last Glacial Maximum. *Science,* 334, 1385-1388, doi: 10.1126/science.1203513.

Sherwood, S. C., S. Bony, and J-D. Dufresne, 2014. Spread in model climate sensitivity traced to atmospheric convective mixing. *Nature*, **505**,37-42, doi:10.1038/nature12829.

Skeie, R. B., T. Berntsen, M. Aldrin, M. Holden, and G. Myhre, 2014. A lower and more constrained estimate of climate sensitivity using updated observations and detailed radiative forcing time series. *Earth System Dynamics*, 5, 139–175.

Spencer, R. W., and W. D. Braswell, 2013. The role of ENSO in global ocean temperature changes during 1955-2011 simulated with a 1D climate model. *Asia-Pacific Journal of Atmospheric Science*, doi:10.1007/s13143-014-0011-z.

Stevens, B., 2015. Rethinking the lower bound on aerosol radiative forcing. Journal of Climate, 28, 4794-4819, doi: 10.1175/JCLI-D-14-00656.1.

Sun, Y., et al., 2014. Impact of mesophyll diffusion on estimated global land CO_2 fertilization. Proceedings of the National Academy of Science, 111, 15774-15779, doi: 10.1073/pnas.1418075111.

van den Bergh, J.C.J.M., and W.J.W. Botzen, 2014. A lower bound to the social cost of CO_2 emissions. *Nature Climate Change*, **4**, 253-258, doi:10.1038/NCLIMATE2135.

Vermeer, M. and S. Rahmstorf, 2009. Global sea level linked to global temperature. *Proceedings of the National Academy of Sciences*, 106, 51, 21527–21532, doi:10.1073/pnas.0907765106.

Voosen, P., 2016. Climate scientists open up their black boxes to scrutiny. *Science*, **354**, 401-402.

Waldhoff, S., Anthoff, D., Rose, S., and R.S.J. Tol, 2011. The marginal damage costs of different greenhouse gases: An application of FUND. Economics, *The Open-Access E-Journal*, No. 2011-43, http://www.economics-ejournal.org/economics/discussionpapers/2011-43

Wigley, T.M.L., et al. MAGICC/SCENGEN v5.3. Model for the Assessment of Greenhouse-gas Induced Climate Change/A Regional Climate Scenario Generator. http://www.cgd.ucar.edu/cas/wigley/magicc/

Zhu, Z., et al., 2016. Greening of the Earth and its drivers. *Nature Climate Change*, doi: 10.1038/nclimate3004

Patrick J. Michaels is the director of the Center for the Study of Science at the Cato Institute. Michaels is a past president of the American Association of State Climatologists and was program chair for the Committee on Applied Climatology of the American Meteorological Society. He was a research professor of Environmental Sciences at University of Virginia for 30 years, and Virginia State Climatologist for 27 years. Michaels was a contributing author and is a reviewer of the United Nations Intergovernmental Panel on Climate Change, which was awarded the Nobel Peace Prize in 2007.

His writing has been published in the major scientific journals such as *Geophysical Research Letters, Journal of Geophysics, Climatic Change, Nature* and *Science* as well as popular serials worldwide. He is the author or editor of seven books on climate and its impact, and he was an author of the climate "paper of the year" awarded by the Association of American Geographers in 2004. He has appeared on most of the worldwide major media.

Michaels holds AB and SM degrees in biological sciences and plant ecology from the University of Chicago, and he received a PhD in ecological climatology from the University of Wisconsin at Madison in 1979.

Slides

AT WHAT COST?
EXAMINING THE SOCIAL COST OF CARBON – SCIENCE SECTION

Patrick J. Michaels
Director,
Center for the Study of Science
Cato Institute

Before the:
Committee on Science, Space, and Technology
Subcommittee on Environment
Subcommittee on Oversight

February 28, 2017

92

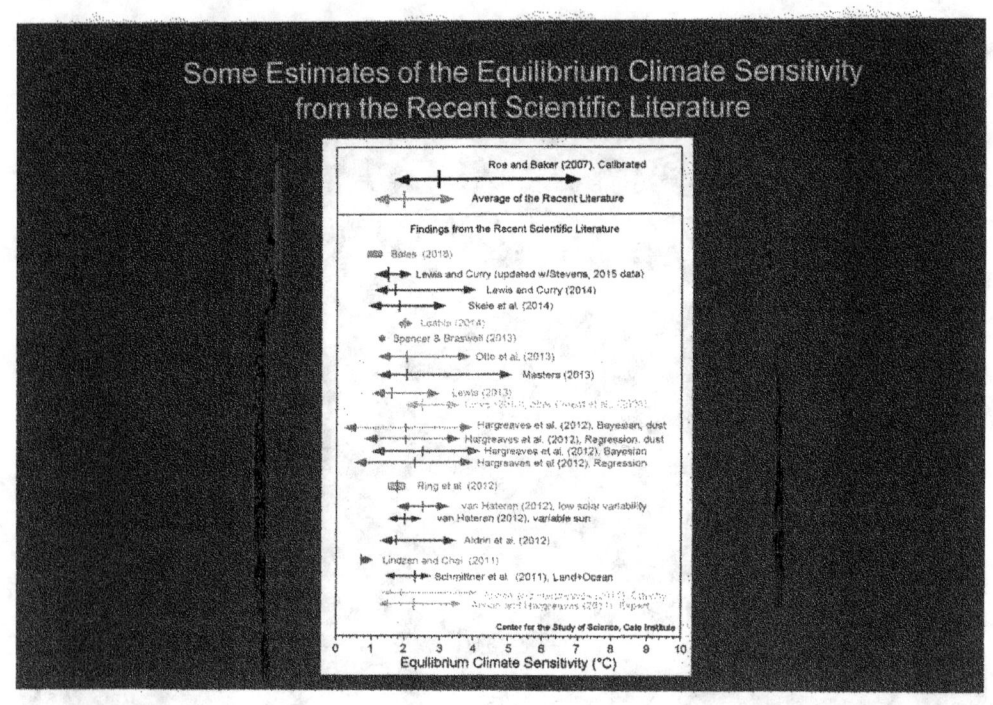

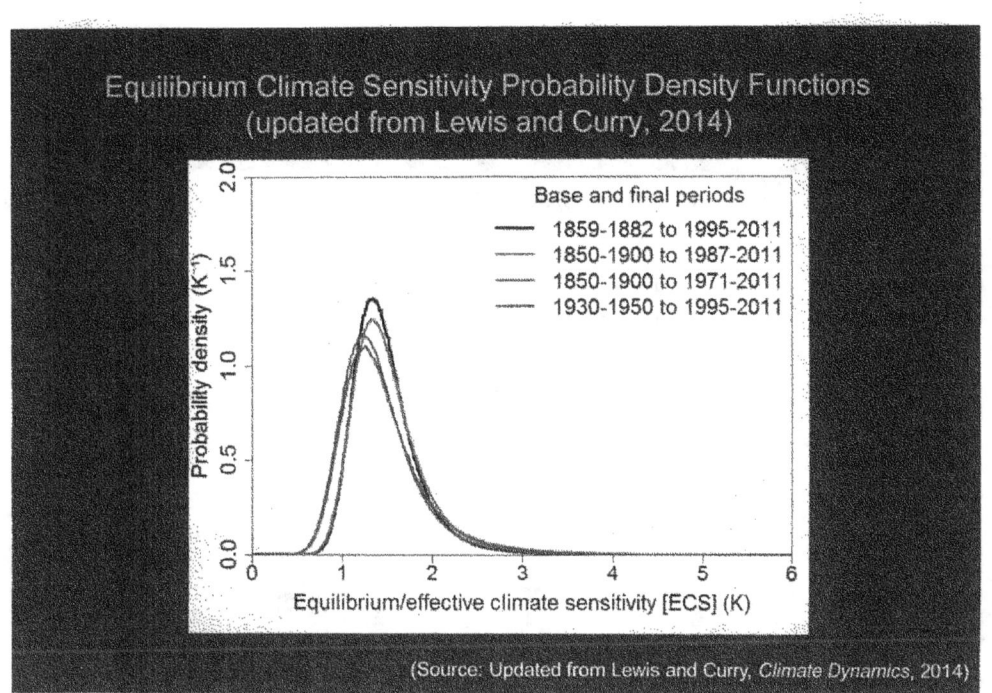

Equilibrium Climate Sensitivity Probability Density Functions
(updated from Lewis and Curry, 2014)

(Source: Updated from Lewis and Curry, *Climate Dynamics*, 2014)

94

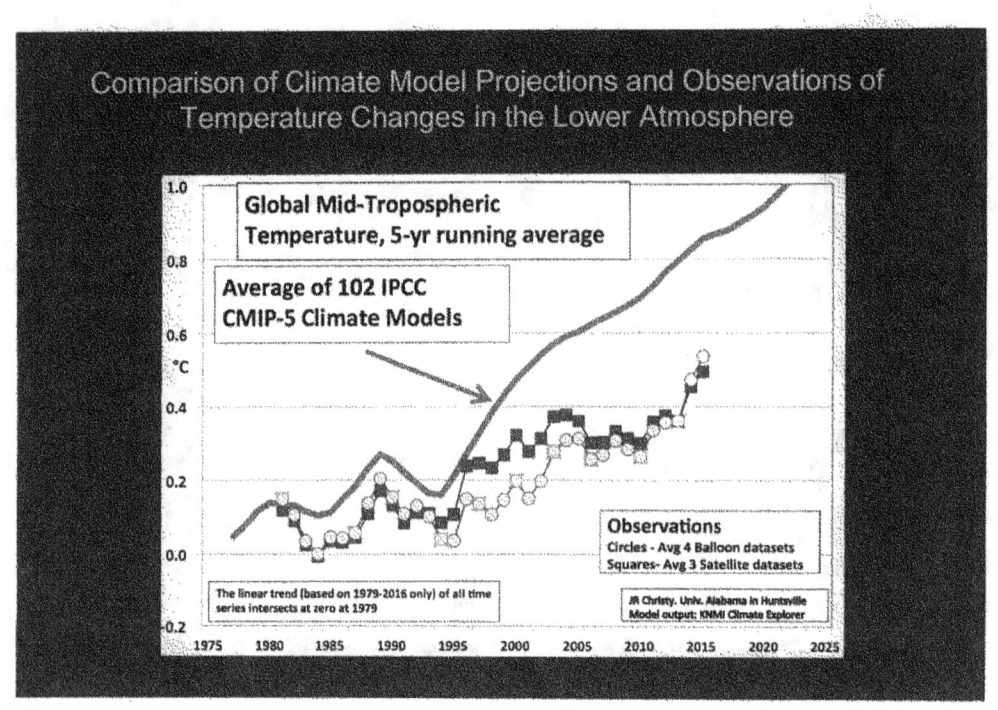

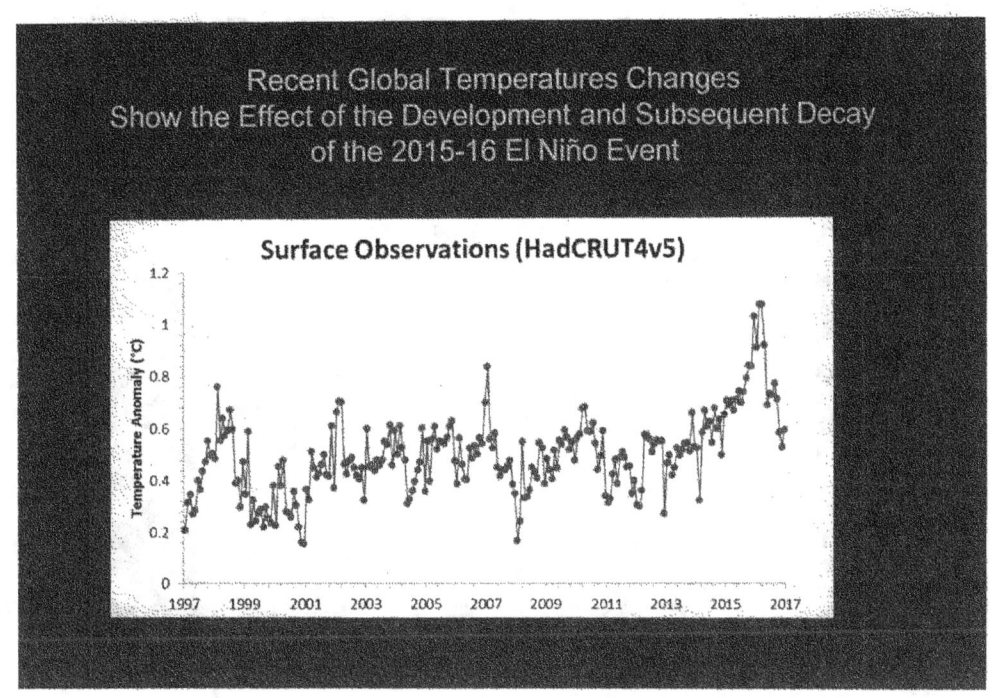

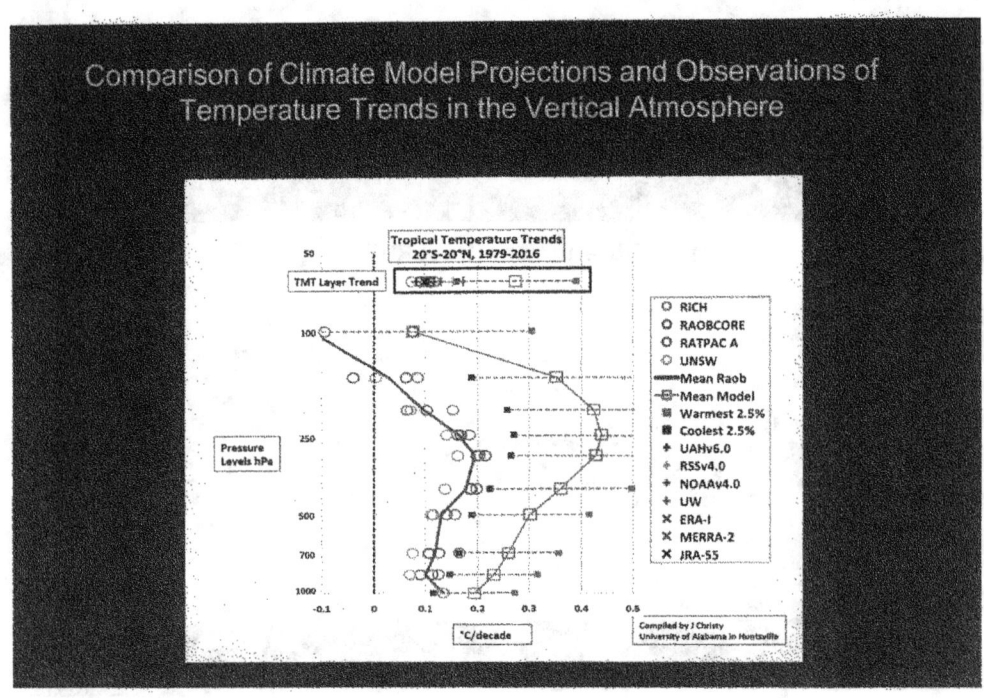

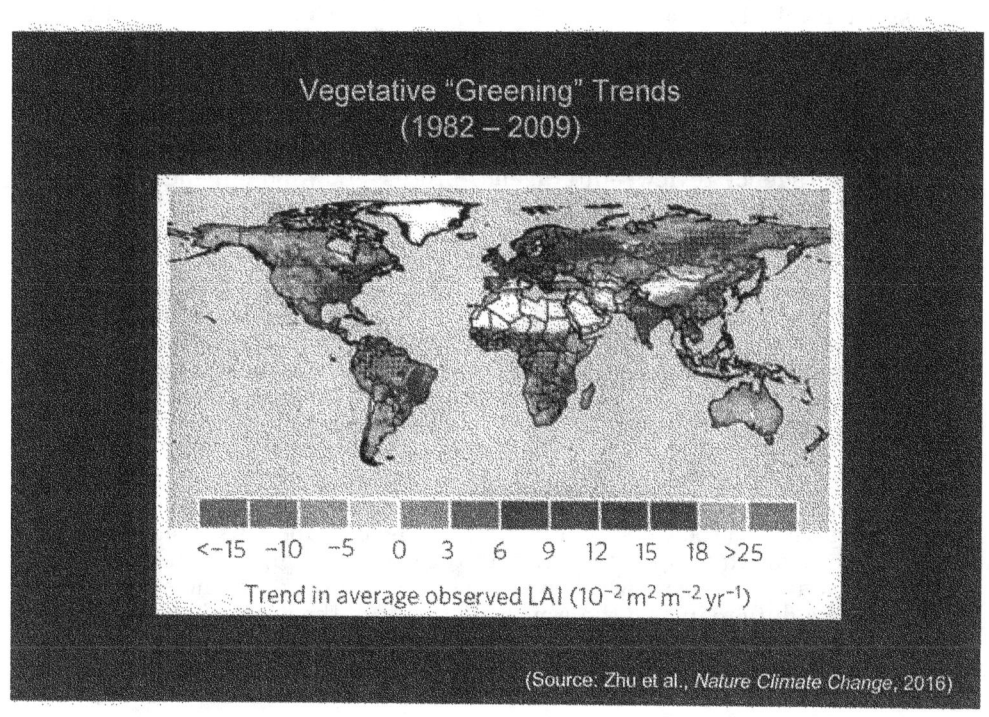

Vegetative "Greening" Trends
(1982 – 2009)

<−15 −10 −5 0 3 6 9 12 15 18 >25

Trend in average observed LAI ($10^{-2}\,m^2\,m^{-2}\,yr^{-1}$)

(Source: Zhu et al., *Nature Climate Change*, 2016)

98

Chairman BIGGS. Thank you. I thank each of the witnesses for their testimony today. The Chair recognizes himself for five minutes.

Dr. Gayer, because the Interagency Working Group ignored Office of Management and Budget guidelines and used a global perspective in determining the SCC, agencies issue regulations with substantial domestic cost based on benefits to non-Americans. In what way does this global perspective method of calculating the SCC potentially mislead the public?

Dr. GAYER. So I'll answer in two parts. First, the—it's—the models that are discussed here today are global models, so it's inordinately difficult for them to come up with domestic estimates, but the interagency review—interagency process did kind of benchmark it to 4 to 14 times differential. So the domestic benefits are 1/4 to 1/14 what the global benefits are based on the models that they used.

And just to clarify, based on comment earlier, I think the approach should be taken, if it is going to use a global measure, it is not a zero-one, right? You could use a domestic measure, you can use a full global measure. Michael referred to straining credibility to assume that there's no global benefits. I think it strains credi- bility to suggest that there's going to be full global benefits, mean- ing if we act, the rest of the world will act instantaneously as well. So to the extent that that doesn't happen and we're using a do- mestic measure, we're overestimating the benefits. And the exam- ple I gave with the Clean Power Plan on the climate benefit side, it's enough to tip the scales that the costs outweigh the benefits.

Chairman BIGGS. And, Dr. Gayer, just to follow up, do agencies have a duty to inform the American public of the domestic benefits in a cost-benefit analysis for federal regulations?

Dr. GAYER. Yes. And those are the OMB guidelines. As I suggested at the end, I think I'd be less kind of worked up over it if they did both, but they should lead with the domestic measure.

Chairman BIGGS. Dr. Dayaratna, what are your biggest concerns about using the SCC in policymaking?

Dr. DAYARATNA. Thank you for the question, Congressman.

So there are a variety of issues with these SCCs, with these IAMs associated with these SCCs. The most fundamental issue is that they are extremely sensitive to very, very reasonable changes in assumptions. As I was referring to, the time—using the time horizon to 300 years, if you shift that to 150 years, which is still unrealistic, you get a drastically different estimate of the SCC.

The discount rate, if you use a seven percent discount rate, as mandated by the OMB, under the FUND model you will get a negative social cost of carbon. And the policy implication there would be that we shouldn't be taxing carbon dioxide emissions but subsidizing it.

Then lastly, with the equilibrium climate sensitivity distributions, there—the ECS distribution used has—was published ten years ago, and it's not even based on empirical research. More up-to-date ECS distributions also will result in a substantially lower and even potentially negative SCC depending on the model that you use.

So with these results all over the map, I do not understand how policymakers can garner any meaningful advice for regulatory policy.

Chairman BIGGS. So just for the laymen who are here, what does the negative value of the SCC kind of connote? I mean, what are we really saying there when we say negative value?

Dr. DAYARATNA. Basically, in a nutshell, in general when people think of SCC they talk about economic damages associated with carbon dioxide emissions. When those damages are negative, that implies that the SCC actually provides benefits. And the result of that is, you know, mostly increased photosynthesis, agriculture, and so forth. But that would suggest that increased CO_2 is actually good for the planet.

Chairman BIGGS. Thank you. Dr. Michaels, in 2015 the Interagency Working Group released a report to the public comments on the determination of the social cost of carbon. You were one of the commenters. What is the significance of having a comment period in the process of developing an SCC? And did the interagency group adopt any of these comments?

Dr. MICHAELS. No, they did not, and many of the comments were well-reasoned based upon recent peer-reviewed literature reinforcing the notion that the sensitivity of temperature that was used in the models was too high. Nature is trying to tell us something. If you look at that satellite image that I showed earlier, you can see that the observed warming in the lower atmosphere is about half of what was predicted. If you look in the tropics, you can see that the observed warming in the vertical in the tropics is about half, actually maybe even a little bit less than what was predicted. All these things point to a consistent story. About twice as much warming was predicted as is going to occur.

Now, the U.N.'s mean sensitivity is 3.2 degrees Celsius. Why don't we settle this out at 1.6, and everybody can go home because we're going to meet the 2 degree guideline with business as usual, declare victory, and let's go on.

Chairman BIGGS. So when the report refused to adopt any of the comments, what does that say to the validity of the report and its objectivity?

Dr. MICHAELS. I would say that they were wedded to a point of view, and I understand. I live in Washington. I understand the pressures in this town. If anyone gave an official answer that this was not a problem, I hate to say we probably wouldn't be chatting here so amiably because nobody would care.

Chairman BIGGS. Thank you, Dr. Michaels.

I recognize Ranking Member Bonamici for questions.

Ms. BONAMICI. Thank you, Mr. Chairman.

Dr. Greenstone, it's been suggested that the Interagency Working Group operated under some sort of veil of secrecy while developing the social cost of carbon. The GAO, in a 2014 report, found that many of the social cost of carbon estimates were developed with input from the public. Now, you and Cass Sunstein convened and led this Interagency Working Group. So did the Interagency Working Group hear concerns that were raised by the other witnesses today, and how were those considered in the process? And can you

also talk about the role that public comment played in the development of the social cost of carbon?

Dr. GREENSTONE. Yes. Thank you, Congresswoman, for letting
me talk about that. So it's probably worth going back in time a little bit. The process—the social cost of carbon was developed in a
very methodical way. The—we convened and led this group that
met many times and it drew expertise from agencies across the Administration. We then first put out—the first put-out was an interim number that was put out for public comment. Then, the final
number of $21 was released in 2010. That has been attached to, I
think, approximately 80 different rules. Public comment was received on that through that process. And then the Administration
later I think in 2013 just put it out for public comment by itself. So
there was tremendous effort to engage the public. There was
tremendous effort to draw expertise from all corners of government,
and it was a highly technical exercise that led to what we perceive
to be and I believe to be describing the frontier of science.

Now, it is possible that some of my fellow witnesses feel spurned,
and I think that's why we often use peer-reviewed literatures to determine what's true and what's not true. And I think just because
their ideas were not judged to be on the frontier does not mean
that the whole process was flawed.

Ms. BONAMICI. Thank you. And, Dr. Greenstone, just because a
comment is made and received but not included does not mean it
was not considered, correct?

Dr. GREENSTONE. Indeed.

Ms. BONAMICI. So, Dr. Greenstone, Dr. Dayaratna said that if the
social cost of carbon was implemented, the country would suffer—
I believe it was—he said disastrous economic consequences, including a loss of jobs and income and an increase in electricity prices.
I'd like you to address what would happen if indeed the Trump Administration has been promoting energy policies without regard to
what the policies may do to the environment. So will you explain
what would happen if we were to roll back regulations designed to
reduce greenhouse gas emissions, if there's an economic price to
pay for undermining the science supporting the social cost of carbon and environmental regulations? And also, how would other
countries like China or India respond if the United States retreats
from or even appears to be retreating from its obligations to address greenhouse gas emissions?

Dr. GREENSTONE. Yes. So I think Dr. Dayaratna is making a very
important point, which is that regulations have costs, so we should
all agree on that and we should—we can move on.

On the flipside of that is that regulations to reduce carbon emissions have benefits, and there's—it's a tradeoff, like many things
in economics. There's no free lunch. And, you know, if we were to
roll back regulations on climate emissions or carbon emissions,
what we would—the world would face and certainly the United
States would face higher temperatures; it would face sea level rise.
It would face a variety of risks that would impose costs on us, our
children, and our grandchildren.

And let me just also return—you also asked, Congresswoman,
about how it would be perceived globally if we reverted to a domestic social cost of carbon, and I think that's an important question

101

because it's not—what—a narrative here about using the domestic damages only I think misses that this is an international problem. And in particular when—as I said in my testimony, when China or India or the EU reduces emissions, it gives great benefits to the citizens of the United States. And to not account for that leverage puts us at risk of higher costs.

And so I think the case for reverting to a domestic-only damage is essentially asking the rest of the world to ramp up their measures.

Ms. BONAMICI. And finally, I know the recent National Academy of Sciences report "Valuing Climate Damages: Updating Estimation of the Social Cost of Carbon" provides some recommendations to the Interagency Working Group on how to improve the methodology. Do the recommendations for an updated estimation undermine the current working group values of the social cost of carbon or would using the Academy's recommendation methodology invalidate existing regulations——

Dr. GREENSTONE. No, the——

Ms. BONAMICI. —that use the——

Dr. GREENSTONE. Thank you for that question. I think rather than have people cherry-pick particular features, which seems to be what's happening today, that they dislike or that they know can change the social cost of carbon, a methodical and scientific approach would be to follow the National Academy of Sciences' recommendations. And indeed, in the Interagency Working Group, we suggested that, as science advanced, the numbers should be updated periodically. And the National Academy of Sciences' report gives a terrific blueprint that would allow for updating.

And, you know, it is true things have advanced since 2009 and 2010. Our understanding about the impacts of climate change, for instance, on human mortality have greatly advanced. And I expected a—following NAS's blueprint would allow for a refresh.

Ms. BONAMICI. Thank you. My time is expired. I yield back. Thank you, Mr. Chairman.

Chairman BIGGS. Thank you.

Dr. Michaels, you wish to respond?

Dr. MICHAELS. Yes, I think it's time to actually take a good, clear look at the effect of policies with regard to the Paris agreement. The EPA uses a model called the Model for the Assessment of Greenhouse Gas-Induced Climate Change. And if you're following, the acronym for that model is MAGIC. It is the standard tool that is used. And you can put in emissions scenarios, climate sensitivity, and come out with a temperature saving as a result of any given policy.

So let's assume a sensitivity that is probably too high, 3 degrees Celsius, and let's reduce United States' emissions to zero right now through the year 2100. The amount of warming that would be prevented would be between 1/10 and 2/10 of a degree Celsius.

Now, let's talk about China and India candidly rather than merely using adjectives and adverbs. The Chinese emission commitments at the Paris agreement are nothing but business as usual. It has long been recognized as their economy matures that their emissions will stabilize around 2030, and that is precisely what they said they would do.

102

The Indian commitment is less than nothing. As economies mature, the amount of CO_2 you emit per unit GDP declines. It's called an increase—or a decrease in emissions intensity. They vowed in Paris to decrease their emissions intensity less, underline less, than the business-as-usual scenario for the country of India.

So really what we do doesn't mean anything to these other countries because they're not doing anything. Thank you.

Chairman BIGGS. I recognize the Chairman of the Oversight Committee, LaHood, for his five minutes of questions.

Mr. LAHOOD. Thank you, Mr. Chairman. I want to thank all the witnesses for being here today and your valuable testimony.

Dr. Gayer, I wanted to ask you, you know, as we try to better understand what Congress and the Trump Administration can do to make agency rulemaking based on more accurate cost-and-benefit information, I do have concerns—I'm not sure whether you're familiar—last year, the Seventh Circuit Court of Appeals, the federal jurisdiction, ruled in a unanimous decision basically against the petitioners, which was an organization by the name of Zero Zone, which is an air-conditioning/heating unit that sued the Department of Energy based on the rulemaking process. And that unanimous decision, their conclusion—and I'm summarizing here—but the DOE conducted a cost-benefit analysis that is within the statutory authority and is supported by substantial evidence. Its methodology and conclusions were not arbitrary or capricious.

And I guess do you have concerns that the court did not criticize this process in this court case?

Dr. GAYER. Thank you for the question.

I feel like I'm in a funny position in many ways. If you lock me and Michael Greenstone in a room, I'm not sure that we'd come up with a very different policy outcome. But I think the process of getting there matters, and I think the regulatory process that was used did involve a lot of assumptions and complexities that I would say lean into arbitrary considerations.

That's not to say I think that the social cost of carbon, this true number out there is negative; I don't. And as I alluded to in my testimony, I do think we should act. I just don't think we should act through kind of the mechanisms of existing statutes through regulations that take a very piecemeal approach and to my mind sort of put a veneer of scientific legitimacy on something that I think is highly, highly uncertain. So that's a longwinded way of me getting at your question, so I apologize.

Yes, I'm concerned when the courts disagree with me to some degree. I don't know what—you know, in the kind of motivation of your question, we are existing in a world where all the action on climate is happening under existing statutes and therefore do—and going through the agencies. And so to the extent that there's too much focus on global, not domestic, I do think it's addressable from Office of Information and Regulatory Affairs and guidance given to the agencies about how they should conduct this.

And I don't want to speak for other people on the panel, but I think there is, you know, a lot of discretion in the choices that were made in how we come up with these numbers. I don't think it was kind of rigged or deception or manipulated. I just think that people disagree. And I think with the new Administration and new OMB and OIRA Director, they can come to a different determination of how much to weigh global versus domestic, as well as issues like discount rates and others, which is a highly difficult, complex, in many ways ethical question.

Mr. LAHOOD. Well, thank you for that.

And just as a follow-up, I think in your opening testimony you went through kind of, you described it as a suite of regulatory policies that we put in place domestically that, you know, have worked fairly well with reducing some of those environmental concerns that people talked about.

In terms of a public policy position here in Congress on what we should do, beyond what you said with working in the Administration, any recommendations for us here in terms of legislation and what we should look at to address this problem?

Dr. GAYER. Well, I think I—you know, I have an iconoclastic view perhaps on this. I first approached this problem because I thought a lot of these regulations—not because I didn't think climate change was a problem but I thought these regulations were much too costly way to go about doing it. And in the analysis that they were using to justify it, I thought they were in some sense making decisions I disagreed with to make them look better than they are.

I testified actually at the House a few years ago. A lot of these regulations are justified not on environmental grounds but because they purport to save consumers money because the underlying thesis there is that consumers are kind of irrational in their consumption decisions, so therefore, you know, the regulator has to come and make the decision for them, which is to my mind kind of a dangerous assumption and shouldn't be used to justify rules.

I have the kind of minority opinion on what should be done, but I think it should go through Congress. I, as I said before, favor kind of a trade for a carbon tax on one side that's revenue-neutral, meaning affording a tax cut and also lighter regulations so that we're not going to the regulatory approach, as we currently have been.

Mr. LAHOOD. Thank you. Those are all my questions.

Chairman BIGGS. Thank you.

I recognize Mr. Beyer for five minutes.

Mr. BEYER. Yes, Thank you, Mr. Chairman, very much.

Thank you all very much for being here. It's fun to have three economists and mathematicians, which we don't often get and a bona fide Ph.D. climatologist. And I want to thank all of you for recognizing that climate change is in fact real. You may have different notions of is it as fast as it was, but this is a great leap forward for the Committee and for America.

And I'm particularly pleased that Dr. Greenstone is the Milton Friedman Professor of Economics, which should make all of my Republican friends very comfortable and happy. And I thank Dr. Gayer for the quote that says that "The social cost of carbon is a conceptually valid and important consideration when devising policies and treaties to address climate change."

Dr. Greenstone, the two things that have come up again and again—in fact, our Chairman's opening statement was, number one, this whole notion of the appropriate discount rate. And you said—you quoted the appropriate discount rate comes down to a judgment about whether climate change involves a substantial risk of being disruptive. Now, the OMB has a circular that says we should use seven percent. Why the decision to ignore the seven percent. Can you tell us simply again why we would choose a number like three percent rather than a 7?

Dr. GREENSTONE. Yes, thank you for your question, Mr. Congressman.

So the discount rate is, as Dr. Gayer mentioned, it's a very tough issue. With respect to climate change or with respect to any public investment, what one would like to do is to use a discount rate where—that reflects an interest rate from the market where the payoffs from that investment are similar to the payoffs from climate mitigation.

And if you think—when you start to think about what climate change might offer, and a lot of it is unknown—as you put it, it could be very disruptive—that would tend to push people to lower discount rates and lower interest rates. And as I said in my testimony, the example of gold is really a good one in the sense that people are willing to hold gold. It has a very low mean rate of return of about three percent, but the reason they're willing to hold it is because it pays off when times are bad. And so if we end up with a bad state of the world with respect to climate, I think that would push us to having—wish we'd used the low discount rate.

It's also worth noting that we're having a somewhat artificial discussion about the 3 and seven percent. Those were set in 2003 when global capital markets looked extraordinarily different than they do today. If we were to instead use what global capital markets are trying to tell us now, the three percent number would very likely be below two percent. That is that's the return on a long—on a long-term government bond. That is a real return. The real return is probably less than two percent to be honest. And there—the seven percent number would also be much lower as well.

So, ultimately, we chose to go with 3 and five percent to reflect the character of the climate problem.

Mr. BEYER. Okay. Great. Thanks.

The second half of that is that there are apparently—according to the majority memo—longstanding OMB guidance to only consider the domestic cost-benefits. And Dr. Gayer I think went on pretty eloquently about, you know, we're considering what's happening around the world, but they're not necessarily affecting their policies. How would you justify the notion of using a global measure of the impacts?

Dr. GREENSTONE. Yes. So actually—so let's establish that I'm not a lawyer, and how nice it was to hear that there was—people were interested in having economists in the room. So that was a surprise. But my understanding of OMB circular A–4, which is what we're talking about, is that it leaves open the option to look at global effects, and that was the path that we drove on. Now, I'm not the legal expert.

The second thing that I want to come back to, and I thought it was very interesting. I saw some light or—between mine and Dr. Gayer's testimony there, which is that I think to do an analysis where—of the benefits of carbon regulations that ignore the leverage from emissions reductions inside the United States, and that leverage comes in the form of emissions reductions in other countries I think is an extraordinarily incomplete analysis. And using the global number is one way and I think a valid way to reflect that leverage.

Mr. BEYER. Thank you. Thank you very much.

Dr. Michaels, congratulations on your Nobel Prize.

Dr. MICHAELS. I didn't say that. That went to the group.

Mr. BEYER. Okay. Well, still.

Dr. MICHAELS. People there, certainly not to the worker bees.

Mr. BEYER. You know, so it was the group that came up with the two percent target I think——

Dr. MICHAELS. Two degree target.

Mr. BEYER. Two degree target, two degree target.

Dr. MICHAELS. That was——

Mr. BEYER. Now you're thinking that we're going to be more like 1.6. Will you be part of the IPCC going forward? And will they come to a 1.6——

Dr. MICHAELS. The——

Mr. BEYER. —consensus in the next couple of years?

Dr. MICHAELS. The numbers that I have always given in my decades of testifying in both the House and the Senate are all within the range of the IPCC consensus so there's nothing new here.

Mr. BEYER. Okay. Great. Thank you very much.

Mr. Chairman, I yield back.

Dr. GREENSTONE. Can I just add one——

Chairman BIGGS. Thank you.

Dr. GREENSTONE. I believe the IPCC——

Chairman BIGGS. Dr. Greenstone, please.

Dr. GREENSTONE. —is from 1.5 to 4 or is it 1.5 to——

Dr. MICHAELS. Yes, I believe 1.6 is in there.

Dr. GREENSTONE. I think you're right at the bottom of the range, yes, but I think the IPCC's consensus is actually—you're right at the bottom of that range.

Chairman BIGGS. Thank you. I recognize——

Dr. GREENSTONE. Just making sure we're all on the same page.

Chairman BIGGS. I recognize Chairman Smith.

Chairman SMITH. Thank you, Mr. Chairman.

Dr. Dayaratna, let me address my first question to you, and it is this: Do you feel that the social cost of carbon is based upon legitimate science or is it based upon arbitrary figures and subjective reasoning?

Dr. DAYARATNA. That's a very interesting question, so thank you, Congressman. In terms of the science, so as, you know, Pat and I both alluded to, the ECS—the equilibrium climate sensitivity distribution that is implemented in these models by the IWG has—was published ten years ago in the journal Science. That is a whole decade ago, and it is not even empirically estimated. It was calibrated to a priori assumptions that the IWG wanted to use regarding climate change.

Now, if you look at the more recent distributions, you will notice significantly lower probabilities of extreme global warming. So what ended up happening was that using this outdated distribution, there was an overstated probability of extreme global warming, and that gets manifested in higher estimates of the SCC. So basically, the SCC estimates were essentially beefed up.

Chairman SMITH. Okay. Thank you.

And, Dr. Michaels, are there benefits to carbon emissions? And if so, should they be factored into the social cost?

Dr. MICHAELS. Well, if you're going to factor costs, you should factor benefits, and the increment just from direct carbon dioxide fertilization for agricultural production is about $3 trillion from 1961 through 2011. But more importantly, the satellite data shows a remarkable greening of the planet Earth, and the illustration that I showed earlier is remarkably reassuring because the massive greenings, the largest greenings are occurring in the margins of the great desert. The Sahelian region in Africa that you and I were taught in school this is desertifying and it will never come back. The tropical rainforest, the lungs of the Earth, have the largest increase in greening on the planet, all brought to you by carbon dioxide. So, yes, you should factor those things in I would think. Thank you.

Chairman SMITH. Okay. Thank you, Dr. Michaels, for that.

One other question. What are some important climate change factors that are not accounted for in the social cost?

Dr. MICHAELS. Oh, God. How many hours do I have to answer that?

Chairman SMITH. How about a minute-and-a-half but——

Dr. MICHAELS. Okay.

Chairman SMITH. Okay.

Dr. MICHAELS. One of the problems is that we spend tremendous amounts of taxpayer money on climate models—

Chairman SMITH. Okay.

Dr. MICHAELS. —and very, very—models for what happens when you increase carbon dioxide and very, very little money on what's called natural climate variability. We know there are these great oscillations in the Atlantic and the Pacific that drive, modulate hurricanes, modulate storm tracks, modulate weather. Those things are not simulated in these climate models, and we need to understand that variability and subtract that out.

I'll close in one second here. The warming of the late 20th century, which began in 1976 and either ended in 1998 or continued—attenuated after 1998 depending upon what we believe—is the same magnitude as the warming of the early 20th century that occurred between 1910 and 20—and 1945. That warming could not have been caused by carbon dioxide.

Chairman SMITH. Right.

Dr. MICHAELS. It means that natural variability is as large as the largest human signal, and yet we only model the human signal. What's wrong with this picture, Congressman?

Chairman SMITH. Yes. Yes. That last point, the last couple points you seldom see covered in the media, but I think they're great points to make.

Final question is should we be using the social cost at all?

Dr. MICHAELS. We should use the social cost of carbon if it's accurately calculated, and I think there's a lot of debate about what we're supposed—what——

Chairman SMITH. Yes.

Dr. MICHAELS. —what it comprises——

Chairman SMITH. Okay.

Dr. MICHAELS. —and what the natural variability component of it is and all this good stuff. We're just not there. It's not ready for prime time, Congressman.

Chairman SMITH. Okay. Thank you, Dr. Michaels.

I yield back, Mr. Chairman.

Chairman BIGGS. Thank you, Mr. Chairman.

I recognize Chairman LaHood.

Mr. LAHOOD. Mr. Chairman, I would just like—I forgot to submit a document for the record from the American Road and Transportation Builders Association regarding our hearing today. I would ask to submit it for the record.

Chairman BIGGS. Without objection.

[The information appears in Appendix II]

Chairman BIGGS. The Chair recognizes Mr. McNerney.

Mr. MCNERNEY. Well, thank you, Mr. Chairman. Thank you for holding the hearing. I thank the witnesses this morning.

Dr. Gayer, I'm very pleased to hear that you support a carbon tax. I think that's the way to go. I'm going to be proposing a carbon tax and benefit package a little bit later, and I hope to get your support on that. Can we follow through with that?

Dr. GAYER. I'd be delighted to, yes.

Mr. MCNERNEY. Very good.

About the domestic versus international impacts, do you believe that the physical impacts of climate change on other nations don't have an impact on our domestic economy or security?

Dr. GAYER. No, I don't believe that, but what I believe is that the actions that we've taken thus far won't lead to reductions matched throughout the entire world. And there are many policies that we have outside of environmental or climate that we—that if other countries did the same thing—and you can think about foreign aid or any number of policies, the world would be a better place and we'd benefit perhaps from it, but we don't take those benefits into account. The regulations where——

Mr. MCNERNEY. So if we drop out of the Clean Power Plan and the Paris agreements, then that's not going to have an impact on China or India or the other countries that are big emitters?

Dr. GAYER. I don't think the Clean Power Plan—well, I don't know that the Clean Power Plan would have an effect. If there's an international agreement and a treaty that is binding, then certainly we should consider the global benefits. Absent that, an EPA regulation I don't think will actually lead to realize the effects throughout the world and certainly not 100 percent throughout the world.

Mr. MCNERNEY. So you think—you do think that climate change is a problem?

Dr. GAYER. Sure. Yes. Yes.

Mr. MCNERNEY. And that the United States should have a leadership role in this issue?

Dr. GAYER. Yes.

Mr. MCNERNEY. Thank you.

Dr. Greenstone, what impact will eliminating the SCC have on current and future environmental protections designed to reduce greenhouse gas emissions? How is that going to affect us if we eliminate that measure?

Dr. GREENSTONE. Thank you for the question. I think it will increase emissions. Of course it would naturally increase emissions in the United States, and that would increase the rate of climate change and global warming. What I think—the point I've been trying to make that I just want to underscore is I think it would also increase emissions in other countries, and so there would be a multiplier effect. And I think it's a mistake to conclude that the Paris agreement did not reflect U.S. leadership and did not reflect that the United States had adopted a robust climate policy. So—you know, let me also add there's I think again surprising agreement on this panel, at least among two of us, that there is climate change. Climate change is real. And there seems to be a little disagreement on the tactics.

You know, our other two witnesses here I think are much more focused on cherry-picking particular features of it, and I think I couldn't agree with them more that updating the assumptions that underlie the social cost of carbon based on the advances in science in the last 7 or eight years is an important thing to do. And indeed, thankfully, the National Academy of Sciences has put out a very clear report on how to go about doing that.

Mr. McNERNEY. Thank you. Now, some of the critics have implied that the SCC is created by the Obama Administration and pushed by environmentalists, but it's my understanding that the Reagan Administration first demanded the Federal Government to do a cost-benefit analysis, and the federal courts required the George W. Bush Administration to monetize these benefits. Is there any other way to do that than using the social cost of carbon?

Dr. GREENSTONE. No. The—really what the courts were requiring that a social cost of carbon be developed. And I think when one thinks back to 2009, what was striking is that even though a ton of CO_2, wherever it's emitted in the U.S. economy has the same impact, you had a complete discordant approach. The department— some departments were treating it as if there were zero costs associated with it, which is, just to be clear, effectively implying that climate change has no negative impacts.

Mr. McNERNEY. Right.

Dr. GREENSTONE. And others were effectively treating it as if it were infinity costs. And so I think landing at the approach and the number that we ended—tried to instill some discipline and coherence across—policy across the government.

Mr. McNERNEY. And so what's the context of how this came about, how this measure came about?

Dr. GREENSTONE. Yes. No, it was—sorry. The Court had required it, and the President had ordered that a number be developed and that—as I mentioned earlier, that used expertise from all branches of government.

Mr. McNERNEY. And this was done in a transparent fashion?

Dr. GREENSTONE. It was done in a transparent fashion. There's been endless opportunities for public comments. It's been at least 80 rules. In addition, it was put out for public comment on its own. Mr.

McNERNEY. Is there some kind of consensus on what parameters to use for this model?

Dr. GREENSTONE. Yes. There was great debate about it, and what—actually what—a rule that I tried to impose when we were leading the process was that we should not be making science—we are, after all, faceless bureaucrats sitting in a room—but instead that our job was to summarize the frontier of science. And I feel that we were quite faithful to that goal.

Mr. MCNERNEY. Thank you. Thank you, Mr. Chairman. I yield back.

Chairman BIGGS. Thank you. Dr. Michaels, you looked like you might want to respond to the assertion of cherry-picking.

Dr. MICHAELS. I would like to—oh, sorry. I would like to respond to the assertion that without the social cost of carbon that our emissions would go up. That's what I call maybe herd reasoning, and I'd like to show you how well herd reasoning works with regard to emissions.

This, which I just happen to be carrying in my backpack, is a shale oil rock. Ten years ago if I said that there were hundreds of years of oil—shale gas rather, which produces half the emissions of carbon dioxide when combusted for power production under our feet, polite company—and we would get it by exploding rocks, polite company would have laughed me out of every Washington cocktail hour. But that's the way people work.

Regulation is not required to create efficiency. Markets are required to create efficiency. This is cheaper than its competitors, and emissions will continue to go down as long as our economy is free for the simple reason that the future belongs to the efficient.

Mr. MCNERNEY. Industry can be relied on to clean itself up. That's basically what you're saying.

Dr. MICHAELS. No, the market can be relied upon to clean up industry.

Mr. MCNERNEY. Mr. Greenstone——

Chairman BIGGS. I'm sorry. The gentleman's time is expired.

Mr. MCNERNEY. I'm sorry.

Chairman BIGGS. The Chair recognizes Mr. Posey.

Mr. POSEY. Thank you, Mr. Chairman.

As policymakers, I think it's important that we all know what we don't know, and therefore, our attempts to predict the future impact of regulations are always speculative and subject to error.

And that being said, it's also true that some predictions are more speculative and uncertain than others. The time between the implementation of a regulation and the onset of any potential benefits is a great example of a factor that makes some forecasts more reliable and others less so. Clearly, the longer the period of time is between the implementation of a rule and the realization of its benefits, the less reliable the analysis of the predicted benefits can be due to the increased likelihood of intervention from unforeseen sources.

My first question is for Dr. Dayaratna. With what I've said in mind, can you give me an idea of the time horizon used in calculating the social cost of carbon? How far into the future are we looking at when we talk about this cost?

Dr. DAYARATNA. The time horizon for computing the social cost of carbon by the IWG is 300 years into the future. And it's interesting that you ask that question, Congressman. Firstly, it's dif-

110

ficult to forecast what the economy will look like, you know, even a couple decades into the future, let alone centuries.

Now, Dr. Michaels had a slide about the temperature projections that John Christy talked about juxtaposing the IPCC's forecast versus satellite and weather balloon data. And I just find it astounding that people would want to use these models to try to make forecasts 300 years into the future when they can't even predict 20.

Mr. POSEY. Yes, we have trouble getting the weather predicted a day ahead of time——

Dr. DAYARATNA. Oh, absolutely.

Mr. POSEY. —oftentimes. So, wow, you're telling me we're basing our regulatory decisions on assumptions about what the world will be like in 300 years. In some ways that's kind of like our Founding Fathers trying to predict and regulate the internet.

Dr. DAYARATNA. Yes. I gave a talk, you know, a couple weeks ago on this topic. You know, John Adams once said America will one day become the greatest empire in the world, and he was right, but yet he'd have no idea what people are doing today using microphones, smartphones, tablets, and so forth. Similarly, we have no idea what things are going to look like that far into the future. And I quite frankly find it's—these models are quite foolish in actually trying to make those types of forecasts.

Mr. POSEY. Okay. Can you describe for us how the social cost of carbon estimates change when you use a more reasonable horizon?

Dr. DAYARATNA. So as I was referring to in my testimony, they change as—by—a reasonable figure is around 25 percent. That figure varies but on average around I think around 25 percent or so, perhaps more.

Mr. POSEY. And given what you just said, do you think it's advisable to continue using the current social cost of carbon estimates in rulemaking proceedings?

Dr. DAYARATNA. Absolutely not. I think these models—you know, they're interesting for academic exercises but they need to be revised to be suitable for regulatory policy.

Mr. POSEY. Do you think in the future the agencies can and should be more forthcoming about the highly speculative nature and variable quality of social cost of carbon estimates?

Dr. DAYARATNA. I tend to think so, yes. To be quite honest, like the sheer fact that they are using an ECS distribution that was—that is ten years old and not even based on empirical research is one thing that is, you know, just not in detail talked about in the IWG's analysis. They did respond to this in the public comments; I will say that. But, yes, there are so many other things out there that they should be more forthcoming about.

And, you know, there was a question that came up earlier about the use of a seven percent discount rate and why it was not used. Quite frankly, here's the reason why I think it wasn't used. Even using the outdated Roe Baker distribution, you still get a negative estimate of the SCC under a seven percent discount rate. That's why it wasn't used.

Mr. POSEY. Okay. There's been a lot of discussion about climate change. Can anyone on the panel give me a date certain, even a year certain that there was absolutely no climate change on this planet since the forming of it?

Dr. DAYARATNA. I believe that the climate has been changing since the planet was first formed.

Mr. POSEY. Any others?

Dr. GREENSTONE. Mr. Congressman, are you talking about—just so we're on the same page, are you talking about climate changing or climate changing due to the release of CO_2? Because I can't quite tell from your question.

Mr. POSEY. I thought I was fairly clear.

Dr. GREENSTONE. Okay. Well, then——

Mr. POSEY. Can you give me any date certain——

Dr. GREENSTONE. The climate has certainly——

Mr. POSEY. —in the history of the Earth that the climate has not changed?

Dr. GREENSTONE. Since the release of CO_2, it has been changing more rapidly.

Dr. MICHAELS. That's not true.

Mr. POSEY. You know, that's speculative, and I didn't ask you to describe a clock. I asked you if you knew what time it was.

Yes, sir, at the end.

Dr. MICHAELS. I think the climate was quite stable about one year before the Big Bang.

Mr. POSEY. Okay. Dr. Dayaratna.

Dr. DAYARATNA. Yes, there was actually a paper published in the Journal of the American Statistical Association last year that looked at tree ring analysis in Tornetrask, Sweden, and they found interestingly using their Bayesian modeling that in 1750 the temperatures there it may have been just as warm if not warmer than they were today in 1750 before, you know, all the things that we—people tend to complain about today.

Chairman BIGGS. Thank you.

Mr. POSEY. Do you have any idea what——

Chairman BIGGS. The gentleman's time is expired.

Mr. POSEY. Any idea what the temperature was pre the last Ice Age when the dinosaurs were roaming the Earth?

Dr. DAYARATNA. Pat actually might have an answer to that.

Dr. MICHAELS. The answer—that's a very good question and I'm glad you asked that. I don't think we have to go pre the last Ice Age. Let's go not to the most recent interglacial but to the penultimate one, the one between what some people call the Illinois glaciation and the Wisconsin glaciation. At the end of the current Ice Age, temperatures got quite a bit warmer than they are now, the beginning of the current interglacial. The beginning of the penultimate one, they were much warmer than they were in our interglacial.

In Greenland, temperatures in summer averaged around 6 degrees Celsius higher than now for 6,000 years. And guess what? The vast majority of the ice on Greenland survived, a heat load that human beings could not put on Greenland if they tried. And then somebody found a skeleton of a bear from 5,000 years before that, and it turns out the DNA sequence was that of a polar bear.

Chairman BIGGS. Thank you. The gentleman's time is expired.

Dr. MICHAELS. Thank you very much.

Mr. Posey. Thank you.

Chairman Biggs. The Chair recognizes my friend from Florida, Mr. Crist.

Mr. Crist. Thank you very much. Good morning.

I was wondering—Dr. Michaels, good morning. How are you?

Dr. Michaels. I'm good.

Mr. Crist. Do you believe that climate change is real?

Dr. Michaels. Of course.

Mr. Crist. Fabulous.

Dr. Michaels. I also believe the sun will rise tomorrow.

Mr. Crist. That's breathtaking.

Dr. Michaels. I know. It's mindboggling.

Mr. Crist. Do you—what would you estimate is the cause of the climate change you believe in?

Dr. Michaels. There are natural causes and there is a human component. You have to understand, the warming of the second— the second warming in the 20th century is accompanied by a cooling of the lower stratosphere. Now, if you change the greenhouse effect, because you change the upwelling flux of infrared radiation, you will warm the lower layers of the atmosphere but you will cool the stratosphere. That's what Karl Popper would call a difficult test of a theory. And indeed, the lower stratosphere cools concurrent with the warming of the troposphere, our neck of the woods. That's a greenhouse signature.

But here's the cool part, Congressman Crist. In 1997, 1998, everybody knows that something happened to warming unless you really jimmy the records and it either slowed down or stopped.

Mr. Crist. Attenuated you said.

Dr. Michaels. Attenuated is a good word because you can attenuate——

Mr. Crist. Sure.

Dr. Michaels. —a lot or you can attenuate a little. But the stratospheric cooling also stopped. Now, if you want me to explain that, I'm going to tell you the three most important words in life.

Mr. Crist. My question is simple. What causes climate change in your estimation?

Dr. Michaels. Lots of things.

Mr. Crist. What's the primary cause?

Dr. Michaels. The fact that we live on a fluid discontinuous earth with long-period oscillations. I mean, the biggest climate change that you and I know of is an ice age oscillation, and I don't think CO_2 is going to be capable of doing that, and those occurred, you know, without human influence. Again, I say the warming of the late 20th century has a greenhouse component because of the stratospheric cooling. By did the stratosphere stop cooling when the surface warming either stopped or attenuated? You know what the answer why that happened is?

Mr. Crist. May I ask another member a question?

Dr. Michaels. No one knows is the answer.

Mr. Crist. Dr. Greenstone, what do you think causes climate change, please?

Dr. Greenstone. I think what we're here to talk about today is the climate change is caused from the release of CO_2——

Mr. Crist. Yes, sir.

113

Dr. GREENSTONE. —of which I think that's a settled issue scientifically. And I can't help but note the contrast here between the concerns of the development of the SCC was an opaque process, which is the claims here. Let me just say the character of the conversation that occurred in those many, many meetings was really quite sober. It was rigorous. It was very scientifically based. And we—at no point did anyone talk about a skeleton of a polar bear as a way to make an argument.

And my own view is that there's a great path forward, and I think there's agreement in the room that there should be a development that the social cost of carbon should be refreshed to reflect the scientific advances that have occurred since 2009 and 2010. And literally, the National Academy of Sciences has outlined a terrific way forward that would also be rigorous, scientifically based, and sober, and I think there's a great opportunity for the Trump Administration to do that.

Mr. CRIST. Thank you. I'm from the Sunshine State, and I'm very proud of that. But having said that, we use less solar energy than New Jersey. And New Jersey's a great place, but it's the Garden State and we're the Sunshine State. And so the point I'm trying to make is if we're going to address climate change and probably the primary cause, which is CO_2, carbon, then wouldn't it be better for us to try to get more of our energy from solar, from sun or wind in order to mitigate the cause?

Dr. GREENSTONE. I think there's a great case for energy markets being in a very unlevel playing field. In particular, the fossil fuels, which involve the release of CO_2, when we go to the gas pump and when we pay our electricity bill, we don't pay for the climate damages that are associated with using them. And if we were to level the playing field so that all sources of energy could compete on equal grounds, it would naturally be the case that there would be a greater reliance on, as you suggested, renewables, probably on nuclear as well, and other low-carbon energy sources.

Mr. CRIST. Thank you, Doctor.

Thank you, Mr. Chairman.

Chairman BIGGS. Thank you.

The Chair recognizes Mr. Rohrabacher from California.

Mr. ROHRABACHER. Thank you very much. And let me just note that I—one of my colleagues suggested that there is a consensus among scientists that global warming is being caused by CO_2 emissions, and let me just note that a consensus may mean 75 percent, it may mean 60 percent, it may mean ten percent and the others don't know for sure. But a consensus is not how you determine whether or not something is scientifically viable. You have to really look to see whether or not it makes sense what people are saying and whether or not, for example, attempts to receive government contracts for research were in some way influencing someone to target their outcome of their research because what I have heard in the last few years is that droughts are caused by global warming and the CO_2 level and then now it's floods are caused by CO_2 level and more tornados. I mean, how many times have we heard that the tornados and the hurricanes are more frequent, but they're not. You know, come to find out they are not more frequent. And all of this coming back to CO_2 and whether or not it is something that we should be concerned about at the level of CO_2.

You know, I drove across the country last year and I saw all these hothouses, and covered-up places, and they were growing all sorts of vegetables. And I went and stopped at several of them and guess what they were pumping into the hothouses? CO_2. Now, why? Because it makes the plants grow better and that means there's more food.

Now, let me ask you this. If we have less CO_2 in the air, does that mean that the plants, and I think we saw something there, that the plants will not grow as robustly if we have less CO_2 in the air? Whoever wants to go into that.

Dr. DAYARATNA. That is actually one of the aspects about the FUND model compared to the other models that is actually incorporated, the feedback from CO_2 into plants and agriculture from CO_2 fertilization. So the other models, the DC. and PAGE model the IWG used, do not account for this type of feedback.

Mr. ROHRABACHER. So that's a great benefit if we have trees that are growing stronger and more trees, more edible plants growing stronger, but that benefit was not calculated into the cost-benefit of other studies?

Dr. DAYARATNA. That benefit was incorporated in the FUND model analysis. Out of the three models used by the IWG, the FUND model actually incorporated that benefit. And, you know, Pat can talk more about this, but there are other benefits in there that could potentially be modeled that, you know, the FUND model doesn't take into account such as, say, aquatic life——

Mr. ROHRABACHER. Okay.

Dr. DAYARATNA. —you know, detailed aspects about vegetation, detailed aspects about agriculture, and so forth that the economy could benefit from.

Mr. ROHRABACHER. Well, let me just put it this way. It's clearer that in the past there were higher levels of CO_2 and great plant life throughout the planet, and we know that and that's very easily discovered in any research. But now, our CO_2 level is looked at as if it's going to be harmful, and let me just say that I don't think that there is a consensus at a high level of percentage, and I think we need to make sure that before we jump into international agreements that it's not just whether it's global benefit or whether it's local benefit. We just have to see whether there's any validity to this concept in the first place so——

Dr. MICHAELS. Congressman, can I offer an observation?

Mr. ROHRABACHER. Please do.

Dr. MICHAELS. I understand that it is thought to be socially responsible to pay for the costs of emission of carbon dioxide, and I also would argue that the fossil fuel-driven societies since 1900 in the developed world have increased their lifespan by 100 percent and their per capita wealth 11-fold. Are we to not also take into account that massive benefit? We should all be dead given our ages in this room if this were 1900, but it is that society that allows us to live.

Chairman BIGGS. Thank you.

Mr. ROHRABACHER. That's a very good point and I'm glad I'm not dead. There you go.

Chairman BIGGS. The gentleman's time is expired. Thank you.

The Chair recognizes Mr. Marshall from Kansas.

Mr. MARSHALL. Thank you, Mr. Chairman. I wish my colleague from Florida was still here. I was going to share with him that we have more sunny days in Kansas than there. Oh, you are still here, Governor. So we have more sunny days in Kansas than Florida, so we look forward to continuing the diversity of energy from Kansas. So thanks for sharing that. I want to acknowledge a good friend of mine, John Francis, who's in the audience, flew all the way in from Great Bend, America, to hear our President speak.

Dr. Gayer, the first question's for you. You mentioned that you would be in favor of a carbon tax being implemented. If we implemented a carbon tax, do you think we could also do away with some of the regulations governing all these carbon producers and just let us pay a tax and be done with it if we can measure it in some way?

Dr. GAYER. Yes. So certainly in many ways that's kind of the thrust of my critique of what's going on in the regulatory sphere. The—and I'm a—Michael Greenstone mentioned leveling the playing field. And so for me the carbon tax is a way to level the playing field, and the regulatory interventions that we've had are a very, very flawed approach to trying to do that. And in my view, the modeling and the global versus domestic are sort of justifications for what I think is a flawed approach.

So, you know, the ultimate trade is a carbon tax in exchange for a tax reduction for more harmful taxes and less regulation and just stick to the pricing.

Mr. MARSHALL. Anybody else want to weigh in on that? Dr. Greenstone, please.

Dr. GREENSTONE. Yes, I think there's a great opportunity for applying a carbon tax at an appropriate level and using—the revenues could be used in a variety of ways. It could be refunded. They could be used to reduce other taxes. And I think they would provide a great opportunity as well for—as a—they could be an excellent substitute for a lot of the regulations that are in place. So there's agreement here.

Dr. MICHAELS. And with regard to the revenue neutrality, I would offer my comment in the form of a question. Do you really expect $3 trillion to walk down K Street unmolested?

Mr. MARSHALL. Okay. I don't know what to say to that. I hope the question wasn't for me.

This social carbon tax is a perfect example of government making the simple complicated. I don't think we'll ever agree. It's a social number; it's a political number. That's what it seems to me and I'm very new to this game.

I guess what I'm more concerned about is, as I watched the Olympics in China and so on and so forth, it would seem to me that whatever measure you use that some of our biggest competitors are producing more of this carbon. And I'm just curious in the big scheme of things in today's world how much carbon is America producing in relationship to China or India, regardless of the social cost we can argue? But what percentage are we now responsible for?

Dr. GREENSTONE. I think this is a rough number. I think historically—I think right now China is producing about 50 percent more
per year than we are. I think we are the second-largest emitter,
larger than the EU, larger than India. I think starting from the Industrial Revolution—someone else might know here—but I'm going
to guess that I think we're responsible for about maybe a quarter
of all emissions.

Mr. MARSHALL. Okay.

Dr. MICHAELS. Yes, but we are becoming more efficient. Our
emissions intensity, which is the amount of CO_2 produced per unit
GDP has dropped more rapidly than pretty much everywhere, and
that didn't happen because of regulations. It happened because of
markets. So if you want efficiency, you would prefer economic——

Mr. MARSHALL. And I want to move on. So I grew up in a small town
between two refineries, oil refineries. So proud that our air in
Kansas is cleaner today than it was when I was growing up and the
waters are cleaner. I want to keep moving in that direction. Back to
my point: manufacturing. I'm trying to figure out why
manufacturing jobs have left Kansas, and one of them is the cost of
energy. Does anybody have any solutions? How do we encourage
China, India, other countries to take leads in this responsibility?
Does anybody have any solutions? Do we tax them or—I don't want
to—does anybody have any solutions on how we encourage them to
get into this game?

Dr. GREENSTONE. I think Dr. Gayer outlined one effective tool,
which would be to have a carbon tax and then have some border
tariff—border tax adjustment so that if people tried to import—so
let's say steel that had carbon embedded in it or carbon was used
to produce it, they would face the same carbon tax that domestic
producers would face.

Mr. MARSHALL. But you would adjust that per country or how
would you figure out—so Europe's doing good, Germany's doing
good, but China's not.

Dr. GREENSTONE. Yes, so you'd have to—there would be some
complexity, I think, but it's imminently doable.

Chairman BIGGS. Thank you. The Chair recognizes Mr. Higgins
from Louisiana.

Mr. HIGGINS. Thank you, Mr. Chairman.

We're here to discuss the real cost of carbon as it's imposed upon
the American people. And it's interesting to note that in discussing
the social cost of carbon I've heard terms like "overwhelming consensus of scientific opinion" and "real science" and yet the measurement standards use a 300-year window to determine actual tax-
ation and cost placed upon the American people. And it's interesting to consider that we're very fortunate that we're not bound
by the science of 300 years ago when we would be discussing ego-
centricity, alchemy, spontaneous generation of life, and the hollow
Earth. And yet scientists tend to speak as if their scientific calculations are absolute and unchallengable. To me, the real overwhelming consensus is that the social cost of carbon is a cost measured not by 300-year windows of manipulated science but the con-
temporary and very real cost of American jobs and American treas-
ure.

So I ask Dr. Dayaratna, you mentioned there are updated equilibrium climate sensitivity distributions. These ECS distributions
quantify the Earth's projected temperature response to a doubling
of carbon dioxide concentrations. As you note, these recent ECS
distributions appear to reflect a lower chance of extreme global
warming in response to increased carbon dioxide concentrations.
Can you explain or can you give us some insight or are you aware
of why the previous Administration, through their Interagency
Working Group failed to update the SCC or any other social cost
of greenhouse gas estimates to reflect these more up-to-date ECS
distributions? If it was not a political decision, then please explain,
what was it?

Dr. DAYARATNA. So the Roe Baker distribution that was published in 2007 is calibrated to a priori assumptions that the IWG
wanted to make regarding global warming based on, you know, a
compiled research discussed by the IPCC. The thing is that—again,
that distribution is calibrated. It is not an empirical distribution.
The percentiles were fitted based to assumptions that the working
group wanted to make. Subsequent ECS distributions are actually
empirically estimated, so they are much more worth considering.

Now, the question regarding why were these new distributions
not included, I think, quite frankly, the reason is that they lower
the estimate of the SCC substantially even if you don't use a seven
percent discount rate, even if you use the assumptions that the
IWG wanted to make regarding 2.5, 3, and five percent. You can
still get a negative SCC using more up-to-date distributions because the fat tail of the Roe Baker distribution has essentially gone
on a diet with the newer more up-to-date ECS distributions, signi-
fying the lower probability of global warming.

Mr. HIGGINS. Dr. Michaels, do you have something to add?

Dr. MICHAELS. I think Kevin is right.

Dr. DAYARATNA. Your mike's not on.

Chairman BIGGS. Please press your mike. Thank you.

Dr. MICHAELS. It was not a lower probability of global warming.
It's a lower probability of high-end global warming——

Dr. DAYARATNA. High-end global—yes.

Dr. MICHAELS. —which is—and that is correct.

Mr. HIGGINS. And with the increase of carbon emissions measured globally, would it not be a reasonable consideration that
greenhouse gas effect would in fact assist the economies of the
earth regarding agricultural production?

Dr. MICHAELS. Well, the effect of carbon dioxide—the direct effect
on plants is well-documented, and the image that I showed at the
end of my presentation, which is a very recent image, documents
the actual greening of much of the Earth, not just the agricultural
component of the Earth. And it's very reassuring to see that the
largest greenings—and they are very, very large—tends to take
place in the margins of the deserts south of the Sahara and in the
northern parts of the tropical rainforest where we were very concerned.

Mr. HIGGINS. Dr. Greenstone, I believe——

Dr. GREENSTONE. Yes.

Mr. HIGGINS. —Mr. Chairman, he has something to add although
I'm out of time.

Chairman BIGGS. Your time is expired. Sorry.

The Chair recognizes Mr. Babin.

Mr. BABIN. Yes, sir. Thank you, Mr. Chairman.

Fascinating testimony. I want to thank everyone for being here, these witnesses.

I represent the 36th District in Texas, which contains the highest concentration of chemical plants and oil refineries of any one district in the entire country. So when the federal government issues carbon regulations based on questionable data and methods, this is of great concern to me because they have a direct and significant impact on my constituents.

And, Dr. Dayaratna, putting things in a perspective from the Industrial Revolution, what association do you see between carbon dioxide emissions and the health of our economy? And along with what some of you folks have already said, obviously it's going to be a drag but I'd like to hear you elaborate a little bit more on that. Dr.

DAYARATNA. Okay. Well, no, that's a very good question, Congressman. So here's the thing. And a lot of people take for granted that energy is a fundamental building block of civilization. So whether it's, you know, powering this room, lighting up our homes, powering our cars and so forth, we all depend on energy.

So when we think about, you know, this whole concept of SCC, the whole goal is to reduce carbon dioxide emissions, and what we end up doing is moving away from the least expensive and most efficient forms of electricity to more expensive and less efficient forms.

So—and the bottom line is economically what we'll notice is that when we go to these so-called lower carbon-emitting, you know, forms of energy, what we would notice is a dramatic change to the economy in the long run. You—you know, a carbon tax, as I was talking about in my testimony, would—in conjunction with the SCC would result in around 400,000 lost jobs on average by 2035, 13 to 20 percent increase in electricity prices, a $2.5 trillion loss in GDP.

Now, on the other hand—and I've also researched this question—if we were to take advantage of the vast shale oil and gas we have in this country, we'd actually see the exact opposite, in fact, even more so in the other direction. We would see a $3.7 trillion increase in GDP, personal income would skyrocket——

Mr. BABIN. Absolutely.

Dr. DAYARATNA. —and, yes, all sorts of things that would benefit the economy.

Mr. BABIN. Absolutely. Thank you so very much for that testimony. And, Dr. Michaels, it's interesting to hear you put a historic and prehistoric context into all this.

Dr. MICHAELS. I've lived that long.

Mr. BABIN. Well, as a student of history, I've read some of the Norse settlements coming from Norway over to Iceland and onto Greenland, and they had settlements there I think from the year 1000 and had trade and routine ships calling on them from Europe for a couple of 300 years. And they were in the process of raising livestock, had hay crops, and then strangely, it had been over 150 years when a ship called on them in the 1500s and none of that community was left. And by that time the climate had cooled off

119

considerably. The hay crops were no longer there. The folks had
disappeared.

So I don't think there was a huge amount of industrialism, carbon dioxide being released into the earth from humans during
those centuries, so if you can kind of address that as well, along
with some of the other——

Dr. MICHAELS. Well, the nature of climate is to change.

Mr. BABIN. Right.

Dr. MICHAELS. It is because we are not a uniform earth. We do
not have a circular orbit. The sun varies and the infrared absorp-
tion of the atmosphere varies, sometimes with human activities.
It's—what do they say? It's complicated, Congressman. And the
problem is in the illustration that I showed, comparing the satellite
and weather balloon observations to the average of the United Nations' 107 computer models shows that it's so complicated that we
haven't gotten close to getting it right and why would you base a
policy upon something that is so blatantly wrong?

Mr. BABIN. Thank you so very much. And the American people
deserve to know the truth here and have sound scientific data, and
that's what this hearing's all about. I want to thank everybody
again for being here, and I'll yield back, Mr. Chairman. Thank you.

Chairman BIGGS. Thank you. The Chair recognizes Mr. Weber
from Texas.

Mr. WEBER. Thank you, Mr. Chairman. I have a question for
each of you all. There seemed to be some discussion about whether
climate change was real and what that meant and the definition,
so here's my question for each of you individually, and we'll start
with you, Dr. Gayer. Would you agree that climate change is
caused by temperature fluctuation?

Dr. GAYER. Yes.

Mr. WEBER. Dr. Dayaratna?

Dr. DAYARATNA. Dayaratna.

Mr. WEBER. Thank you.

Dr. DAYARATNA. The question is would I agree that climate
change is caused by——

Mr. WEBER. Is caused by temperature fluctuation?

Dr. DAYARATNA. Yes.

Mr. WEBER. How about you, Dr. Greenstone?

Dr. GREENSTONE. It's a function of temperature variation. It's
also a function of CO_2 emissions.

Mr. WEBER. Temperature variation is a good one, too. I didn't
mention CO_2. I'll come back to you. Dr. Michaels, would you agree
climate change is caused by temperature fluctuation?

Dr. MICHAELS. It is the contrast—oh, sorry. It is the contrast in
temperature between the surface and the upper atmosphere that
derives—drives most of the precipitation mechanisms on Earth, so
the answer would be if that changes, yes.

Mr. WEBER. Okay. I'll take that as a yes.

Dr. Gayer, would you agree that temperatures fluctuate when
seasons change?

Dr. GAYER. Yes.

Mr. WEBER. Doctor?

Dr. DAYARATNA. Yes.

Mr. WEBER. Dr. Greenstone?

120

Dr. GREENSTONE. Yes. I also think they fluctuate——
Mr. WEBER. Dr. Michaels?
Dr. GREENSTONE. —from CO_2 emissions.
Dr. MICHAELS. Me four.
Mr. WEBER. Okay. Dr. Gayer, back to you. Would you believe—would you agree that temperatures fluctuate in historical, global, cyclical fashion? In other words, we have historical evidence that temperatures changed up or down historically.
Dr. GAYER. Yes, I don't know cyclical necessarily but yes——
Mr. WEBER. Well, okay.
Dr. GAYER. —it's gone up, it's gone down.
Mr. WEBER. I'll give you that. How about—
Dr. DAYARATNA. Yes.
Mr. WEBER. —you, Doctor?
Dr. DAYARATNA. Yes.
Mr. WEBER. Dr. Greenstone, would you agree with that?
Dr. GREENSTONE. Yes. I also think that it varies——
Mr. WEBER. It's just a yes or no.
Dr. GREENSTONE. —because of CO_2 emissions——
Mr. WEBER. Dr. Michaels?
Dr. MICHAELS. I will use the weasel word quasi-cyclical.
Mr. WEBER. Got you. Okay. Now, would you agree also, Doctors, that the temperatures actually fluctuate more when seasons change? Obviously, they go up drastically in Texas to 100, 110 in the desert area sometimes or they go way down below, so when it changes from fall to winter, for example, temperatures fluctuate wildly. Would you agree with that, Dr. Gayer?
Dr. GAYER. I'm confused by the question because I thought that was the previous question.
Mr. WEBER. Would you agree that temperatures fluctuate more when seasons change than they just do from week to week, for example?
Dr. GAYER. Yes.
Mr. WEBER. Okay. Dr. Dayaratna?
Dr. DAYARATNA. Yes.
Mr. WEBER. Okay. Dr. Greenstone, minus the COT component—CO_2?
Dr. GREENSTONE. I think CO_2's important in terms of temperature. I also think seasons are important in terms of temperature.
Mr. WEBER. Dr. Michaels, would you agree they fluctuate more wildly when seasons change?
Dr. MICHAELS. Yes, sir.
Mr. WEBER. Great. So we know that the temperatures fluctuate when seasons change. Now, we're talking about a carbon tax. And so if you go back to where we're going to charge carbon tax for people on industry or countries, let's say, that have industry, are we going to take into account when their seasons change because now they're using more electricity when it's hot or more electricity when it's cold? Do you take that into account at all in the proposed carbon tax?
Dr. GAYER. I'm not——
Mr. WEBER. Do they get a credit when——
Dr. GAYER. No, the——
Mr. WEBER. —they have a mild season.

Dr. GAYER. The goal of the carbon tax is to include a price into the energy decision. So certainly when they use more energy, the tax will go higher and the——

Mr. WEBER. Okay. So they could put——

Dr. GAYER. If you're tying it to like tax reduction somewhere else, the revenue would go higher then.

Mr. WEBER. They do get a credit when it's mild. I got you. Okay. Now, what happens when those countries have a tremendous catastrophe, whether it's a huge hurricane or a huge cyclone, tsunami, or you name it, and they are really hard hit and they have to have more energy production to rebuild their country, do they then get a tax credit to be able to go back and rebuild their country or do we punish them more because now they're using more energy to rebuild?

Dr. GREENSTONE. Can I ask a clarifying——

Mr. WEBER. No, I'm asking him first, Dr. Gayer.

Dr. GAYER. I don't—I didn't—I don't understand the tax——they level—you level—the tax increases the price of energy, yes.

Mr. WEBER. So no matter what happens in a country, if they have a huge catastrophe and they have to use a lot of energy to rebuild their country, they don't get a break? They're just going to pay more carbon tax at that point?

Dr. GAYER. Yes, that's the nature of a tax.

Mr. WEBER. Doctor, do you agree with that?

Dr. DAYARATNA. That—I've never put together a carbon tax proposal myself——

Mr. WEBER. I'm just——

Dr. DAYARATNA. —but in principle, the—yes, that seems to be what——

Mr. WEBER. That's what's going to happen.

Dr. DAYARATNA. —we would want to do, yes.

Mr. WEBER. Dr. Greenstone, do you agree that's going to happen, they use more energy, more carbon to rebuild their country and they're going to get taxed on it?

Dr. GREENSTONE. I just want to clarify if we're talking about a hurricane disaster that's due to CO_2 accumulation in the atmosphere or just one that has——

Mr. WEBER. I'm talking about the tax once they have a disaster. Do you know where I'm going, Dr. Michaels? Can you see what I'm asking here?

Dr. MICHAELS. Yes. I believe that you are drawing the analogy to what happened to this House when it passed cap-and-trade in 2009.

Mr. WEBER. Well, that was a catastrophe all right but——

Dr. MICHAELS. Correct.

Mr. WEBER. So here's the point I'm making. Now, suppose an industry comes along and they develop a process of capturing CO_2 and putting it underground. Do we revoke the carbon tax?

Dr. GAYER. No, you credit it. That's the—that——

Mr. WEBER. Credit it?

Dr. GAYER. That's—and that's one of the nice incentives of having a tax because it incentivizes those kind of technological improvements.

Mr. WEBER. Okay.

Dr. GREENSTONE. In fact, it would be terrific. It would provide a market incentive——
Mr. WEBER. Okay.
Dr. GREENSTONE. —to engage for people to find ways to re- duce——

Mr. WEBER. Okay.
Dr. GREENSTONE. —CO_2 in the atmosphere.
Mr. WEBER. Okay. Well, let me just——
Dr. GAYER. And be more resilient going forward.
Mr. WEBER. —add for the record, Mr. Chairman, and I'm done that in my district we have the largest carbon capture storage unit, Air Products in—over in Jefferson County in the country. So just interesting food for thought where we're headed with this idea.
Mr. Chairman, I yield back.
Chairman BIGGS. Thanks. I thank the witnesses for their valuable testimony and the members for their questions. The record will remain open for two weeks for additional comments and written questions from members. This hearing is adjourned.
[Whereupon, at 12:03 p.m., the Subcommittees were adjourned.]

Appendix I

ANSWERS TO POST-HEARING QUESTIONS

ANSWERS TO POST-HEARING QUESTIONS

Responses by Dr. Kevin Dayaratna, PhD
<u>At What Cost? Examining the Social Cost of Carbon</u>
Questions for the Record

Kevin D. Dayaratna, Ph.D.
Senior Statistician and Research Programmer
The Heritage Foundation

1. It is difficult to understand how an economy, as well as a planet's climate, will evolve decades, let alone centuries, into the future. For example, many aspects of society that we take for granted today such as GPS technology, cellular phones, fax machines, and tablet computers among many others would have been mere science fiction a century ago. Regarding climate, below is a chart presented by Dr. John Christy during his testimony before this the House Science and Technology Committee, juxtaposing IPCC forecast against actual satellite and weather balloon data. As the chart illustrates, the standard IPCC models perform very poorly at making forecasts into the future:

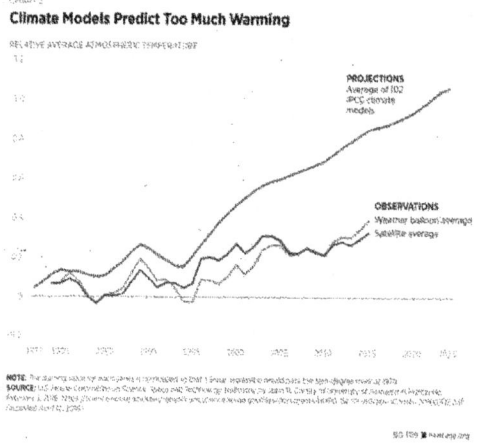

(from Kreutzer et al, "The State of Climate Science: No Justification for Extreme Policies" April 2016.)

One can easily see the difficulty (and gross over-predictions) of the climate models versus actual temperature observations. In my own analysis, I have extracted sea level rise projections from the DICE model:

Sea Level Rise (DICE)

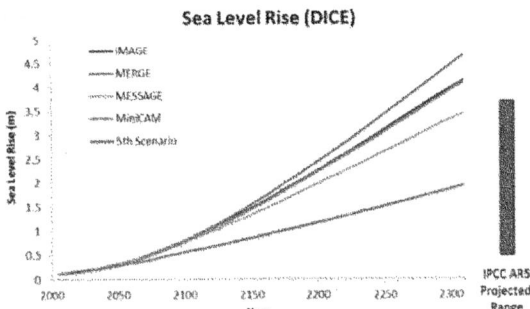

Note that three of the five projections made by the DICE model even exceed the upper end of sea level rise made by the IPCC in their Fifth Assessment report. It is difficult to make forecasts three centuries into the future, and regulators and bureaucrats can easily manipulate these forecasts to conform to a particular policy agenda. I thus believe that these 300-year projections are not only difficult to generate but can easily lend the models that they are used for to user-selected manipulations.

2. In my research, I have examined the DICE and FUND models. Of these models, the FUND model actually allows potential benefits of carbon dioxide emissions, while the DICE model does not. In fact, under some very reasonable assumptions, the FUND model suggests that these benefits may actually exceed the costs of carbon dioxide emissions. The distributional properties regarding this negativity are discussed in detail in my research and are referenced in my testimony.

3. We noticed in our analysis that even using the outdated Roe Baker ECS distribution that the IWG assumes, the FUND model produced a negative estimate of the SCC under a 7% discount rate. I believe that this negativity is large reason why a 7% discount rate was ignored by the IWG despite the fact that its inclusion is mandated in OMB Circular A-4; using such a rate would have required the IWG to present these negative estimates in its report.

4. A variety of the IAMs' components lack sufficient empirical justification. For example, the equilibrium climate sensitivity distribution (Roe Baker, 2007) used by the IWG is calibrated to a priori assumptions regarding the climate, and hence its parameterization is not based on empirical data. Additionally, as I have noted above, assumptions regarding sea level rise made in the DICE model even exceed estimates made by the IPCC. Moreover, the damage functions behind the IAMs themselves are arbitrary, devoid of legitimate empirical justification. Therefore, I believe that the IAMs have not been validated at a level that would suffice for them to be used in regulatory policy.

5. I have been quite easily able to download and install the Model for the Assessment of Greenhouse Gas Induced Climate Change (MAGICC). To obtain the IAMs, on the other hand, I had to directly contact the EPA. After doing so, I was able to obtain only two of the three

models - the DICE and the FUND model. The EPA would not give us the PAGE model, asking us to contact the model's author, Chris Hope, for the codes. Hope said that he would only give us the codes if we allowed him to be a co-author of any research we publish. We felt that this stipulation precluded us from being able to do independent analysis of the model. As a result, based on my experiences, these IAMs do not seem to be as freely available as the MAGICC model.

Additionally, from what I understand, the MAGICC model was the primary model used for environmental regulatory policy before these IAMs. Since the MAGICC model had been so heavily relied upon, it would have been useful for the IWG to compare their analysis of the climate to MAGICC. I have not, however, seen such a comparison made.

A third difference between the MAGICC model and the IAMs is that the MAGICC model provides choices of assumptions including climate sensitivity as well as other forcing criteria. The computer codes for the IAMs, on the other hand, adhere to assumptions made by the IWG, and to alter these assumptions, I had to edit the codes myself.

6. Using the MAGICC model, we simulated a hypothetical situation of eliminating carbon dioxide emissions from the United States completely. We found that, using the most extreme assumptions regarding climate sensitivity available in the model, the impact of such a scenario would result in less than 0.2 degree Celsius temperature mitigation and less than 2 cm of sea level rise reduction. Our analysis thus suggests that the United States' impact on the climate with respect to the rest of the world is negligible.

7. It is quite frankly disturbing that those associated with the IWG were part of the NAS report. This association suggests that politics is guiding science, not vice versa.

8. I have not seen any meaningful references to benefits of carbon dioxide emissions in any of the IWG's analyses. For example, the IWG could have, but instead avoided, reporting the probability of a negative social cost of carbon, as I have done in my research. This glaring omission is exactly what I was referring to in my testimony - It seems as if the IWG was deliberately trying to inflate its estimates of the SCC to justify a regulatory agenda.

9. Advocates of a regulatory agenda will always try to defend these models in court. The IWG claimed that it "responded" to comments; however, I noticed that its responses to our comments were just acknowledgements of statements we made and were quickly disregarded. As I have discussed in my testimony as well as in my research, these models can be manipulated by a priori assumptions to obtain any result the user desires. I therefore fundamentally disagree with the court's decision and believe that, at this point, these IAMs are not suitable for guiding regulatory policy.

Responses by Dr. Michael Greenstone, PhD
House Committee on Science and Technology
Testimony on the Social Cost of Carbon – Follow-Up Questions
Michael Greenstone

Question 1
During the hearing, Dr. Michaels conflated the reduction of the U.S.'s carbon intensity in recent years with markets driving the country and industries toward greater efficiency. He claimed that regulations are not required to create efficiency, but that efficiency is the result of the markets.

a) *Can the reduction in the U.S.'s carbon intensity in recent years be solely attributed to markets pushing various sectors towards greater efficiency?*

The recent reduction in carbon intensity across the United States has largely been driven by a shift in electricity production from coal to natural gas due to recent technological advances in hydraulic fracturing, allowing for inexpensive recovery of natural gas (and petroleum).[i] From 2010 to 2015, US carbon intensity fell by 10%, and the electric power industry accounted for 77% of this reduction.[ii] As it happens, natural gas is less carbon intensive than coal, so this shift also caused a decline in carbon intensity – but this was not the catalyst, only a side-effect. The market mechanism causing this shift from coal to gas is the minimization of private cost, and the reductions in carbon intensity were largely an unintended consequence.

It is important to note that when ignoring damages from pollution and greenhouse gas emissions, the least costly way of producing energy is not necessarily the least carbon intensive. Natural gas is still much more carbon intensive than solar, wind, hydro, and nuclear technologies, all of which are more expensive. That is to say, the market has largely chosen natural gas because its private costs of production are the cheapest.

b) *What role did regulations play in this reduction?*

I believe policy has created an additional set of incentives to abate carbon. For example, from 2005 to 2015, there has been a large reduction in carbon intensity in the transportation sector, largely due to EPA and DOT fuel economy regulations. In this period, per-mile CO_2 emissions fell by 22%, while fuel economy improved from 20 to 25 miles/gallon largely due to transportation fuel economy standards.[iii]

In the power sector, a host of different regulations have aided the reduction of carbon intensity. For example, 29 states have enacted renewable portfolio standards (RPS), which mandate that a certain percentage of electricity must be produced by renewable technologies (with the definition of renewable varying across states). From 2000 to 2013 the percent of total electricity driven by renewables grew from 7% to 13%, and 60% of that growth has been attributed to RPS standards.[iv,v] As a result, in 2013 alone, RPS was responsible for a 59 million tonne reduction in CO_2 emissions nationally.[vi]

In addition to these two, there have been a host of other regulations that have contributed to the recent decline in carbon intensity: California's AB32 rule, which implements a cap-and-trade on GHG emissions in the state; and investment and production tax credits for renewables, to name a few.

In summary, the reduction in US carbon intensity has occurred largely because of market forces that choose energy sources with the lowest private costs and, to a lesser extent, because of environmental and energy regulation. The environmental consequences of these market forces are largely unintended consequences that cannot be counted on to deliver emissions reductions. Policies that price pollution and greenhouse gas emissions would level the playing field and cause markets to choose the cheapest energy sources, after accounting for their full costs.

Question 2

Despite scientific consensus supporting the negative impacts of climate change, we occasionally hear from our colleagues across the aisle that climate change may actually have some beneficial effects. Specifically during the hearing, there was discussion about the positive economic impacts of the greening of the earth due to an increase in atmospheric carbon dioxide.

a) *Is there any additional clarification you would like to add to this discussion?*

I agree that increased fertilization and vegetation ("greening") in some parts of the world is among the many impacts of increasing atmospheric CO_2. As discussed at the hearing, these impacts would be positive for plant growth. However, I'd like to add a clarifying point to this discussion, namely that this is just one of the effects of the accumulation of CO_2 in the atmosphere, and it is important to get a holistic, not a partial picture.

Whatever the magnitude of positive terrestrial greening, the best evidence indicates that CO_2 accumulation's net impacts will be negative at the levels of CO_2 that we are rapidly approaching (and possibly already reached). The negative impacts include higher global mortality and increased energy demand due to higher incidence of very hot days; large economic damages due to rising sea levels; and the various costs from stronger hurricanes.[vii] They also include large adverse effects on agriculture due to increased aridity, as well as increased incidence of pests and weeds.[viii] These impacts are especially relevant in the context of the CO_2 fertilization debate, because estimates of the benefits from fertilization are overstated if they do not take into account how global warming adversely affects agriculture.

b) *How are those claims of the beneficial effects of carbon dioxide emissions captured by the social cost of carbon?*

The IWG already incorporates the beneficial effects of CO_2 into their estimates of the social cost of carbon. The FUND model, used by the IWG in conjunction with the DICE and PAGE models to estimate the SCC, incorporates estimates of agricultural benefits due to increased fertilization in its estimation. In fact, the FUND model predicts small net positive effects of climate change at low levels of temperature change, so it is evident that the beneficial effects of warming are included in the social cost of carbon. These net positive effects are not evident in the other two models.

More broadly, the key point is that the net effect of climate change is the relevant parameter. The best prediction across the three models that underlie the SCC is that climate change will impose costs at virtually all increases in temperature. Further, these costs grow nonlinearly with the increase in temperature.

Question 3

A factsheet published by the Institute for Policy Integrity states that corporations, not just states and federal agencies, use a value for the cost of carbon. "Many major companies also quantify the cost of carbon pollution in their financial planning." Companies such as Microsoft, GE, Disney, and Walmart to just name a few. Even the Exxon Mobil Corporation uses $80 for a metric ton of carbon dioxide emissions in 2040.

a) *Why are businesses using their own social cost of carbon to assist their planning?*

This issue is one that neither I nor the broader economics literature have spent much time researching. I'm afraid that I would need to study these trends further to be able to offer a meaningful reply.

b) *Why should we take a cue from industry leaders and continue using the social cost of carbon as a metric for policymaking?*

I'm afraid that I would need to study these trends further to be able to offer a meaningful reply.

c) *What is your reaction to Exxon using a value more than twice as high for carbon emissions than the current social cost of carbon?*

With regards to Exxon Mobil's large SCC of $80, there is an important clarification to be made here. Exxon uses $80 for 2040, and the USG uses around $40 for *2015*. With a 3% discount rate, the USG SCC is $69 in 2040 and with 2.5% it is $96 in 2040.[xi] Comparing Exxon 2040 to USG 2040 is more apples-to-apples, and in doing so, $80 seems like it was pegged to the USG social cost of carbon.

Question 4

The science supporting climate change has no political affiliation and neither do the impacts of a changing climate discriminate based on party lines. The Risky Business project, a project co-chaired by Michael Bloomberg, Tom Steyer, and Henry Paulson, outlines staggering economic consequences of continuing business as usual when it comes to dealing with climate change. For example, the Project estimates that by 2050, between $66 and $106 billion worth of existing coastal property will likely be below sea level nationwide.

    a) *What happens if we start doing less?*

Any policy that reduces mitigation of carbon emissions will increase the probability of imposing high-cost scenarios on future generations. Furthermore, curbing US climate is likely to encourage other countries to reduce their climate mitigation efforts. As evidence of the United States' central role, other countries increased their pledges to cut emissions as part of the Paris Agreement after the U.S. and China reached an agreement to increase their climate mitigation efforts.

    b) *How do you respond to those who claim these types of economic consequences are inflated or imaginary?*

Ultimately, society needs to balance the costs to our economy of mitigating climate change today with climate damages in the future. Wishing that we did not face this trade-off will not make it go away. We cannot know the costs of climate change with certainty but trying to make them go away by pretending they don't exist is choosing to shift all of the costs of our choices to our children and their children and so on.

131

References

[i] U.S. Energy Information Administration. *Hydraulically Fractured Wells Provide two-thirds of U.S. Natural Gas Production,* by Jack Perrin and Troy Cook. https://www.eia.gov/todayinenergy/detail.php?id=26112

[ii] U.S. Energy Information Administration. *March 2017 Monthly Energy Review.* DOE/EIA-0035(2017/3) (Washington, DC, 2017). https://www.eia.gov/totalenergy/data/monthly/pdf/mer.pdf.

[iii] U.S. Environmental Protection Agency. *Light-Duty Automotive Technology, Carbon Dioxide Emissions, and Fuel Economy Trends: 1975 through 2016.* EPA-420-R-16-010 (Washington, DC, 2016), https://www.epa.gov/sites/production/files/2016-11/documents/420r16010.pdf.

[iv] U.S. Energy Information Administration. *Net Generation for All Sectors, Monthly.* (Washington, DC, 2017) https://www.eia.gov/electricity/data/browser/#/topic/0?sec=g&geo=g&fuel=vvg&agg=2,0,1 (accessed March 2017).

[v] Lawrence Berkeley National Laboratory. *U.S. Renewables Portfolio Standards: 2016 Annual Status Report*, by Galen Barbose. LBNL-1005057 (Berkeley, CA, 2016). https://emp.lbl.gov/sites/all/files/lbnl-1005057.pdf.

[vi] Lawrence Berkeley National Laboratory. *A Retrospective Analysis of the Benefits and Impacts of U.S. Renewable Portfolio Standards,* by Ryan Wiser, Galen Barbose, Jenny Heeter, Trieu Mai, Lori Bird, Mark Bolinger, Alberta Carpenter, Garvin Heath, David Keyser, Jordan Macknick, Andrew Mills, and Dev Millstein. NREL/TP-6A20-65005 (Berkeley, California, 2016). http://www.nrel.gov/docs/fy16osti/65005.pdf

[vii] World Bank. *Turn Down the Heat: Confronting the New Climate Normal.* (Washington, DC, 2016). https://openknowledge.worldbank.org/handle/10986/20595.

[viii] Environmental Defense Fund, Institute for Policy Integrity, Union of Concerned Scientists, and National Resource Defense Council. "Comments on the Department of Energy's Use of the Social Cost of Carbon." (Washington, DC, 2016). http://policyintegrity.org/documents/Joint_Comments_to_DOE_Nov2016.pdf

[xi] U.S. Environmental Protection Agency. *EPA Fact Sheet: Social Cost of Carbon.* (Washington, DC, 2015). https://www3.epa.gov/climatechange/Downloads/EPAactivities/social-cost-carbon.pdf. Dollar amounts converted to 2015 dollars using CPI.

Appendix II

134

STATEMENT SUBMITTED BY FULL COMMITTEE
RANKING MEMBER EDDIE BERNICE JOHNSON

OPENING STATEMENT

Ranking Member Eddie Bernice Johnson (D-TX)
House Committee on Science, Space, & Technology
Joint Subcommittee on Environment and Oversight hearing,
"At What Cost? Examining the Social Cost of Carbon"
Tuesday, February 28, 2017

Thank you Mr. Chairman.

In 2014 alone the U.S. released nearly 7 billion metric tons of carbon dioxide into the atmosphere. The Social Cost of Carbon is an estimate of the economic damages caused from the release of a single metric ton of carbon dioxide (CO2) emissions.

Put simply, the Social Cost of Carbon attempts to quantify the economic consequences of our fossil fuel related actions. And let me be clear, our actions <u>do</u> have consequences. Just as our health is impacted by what we put into our bodies, the planet is affected by the chemicals we release into the environment. Denying this reality does not erase the fact that this is true. These acts come with financial costs and social consequences to our environment, to public health and to our economy.

Unfortunately, the Majority too often denies these truths and continues to say, <u>**no,**</u> to basic facts.

They say, <u>**no**</u>, human-influenced climate change is not occurring; despite the enormity of the scientific evidence.

They appear to have <u>**no**</u> concerns about the impact on public health of the release of toxic chemicals into the environment by oil, gas and mining industries.

And some of them believe the federal government should have virtually <u>**no**</u> role in helping to inform the public of these dangers, or, to help protect them by holding industry accountable for their actions.

Americans understand, that "no" is not the answer. **No** does not erase the mountains of scientific evidence that point to climate change. **No** does not diminish the ethical and legal responsibility of private industry to <u>not</u> poison the public by producing and releasing toxic chemicals into our neighborhoods, communities or the atmosphere, or, simply denying the reality of their actions and the resulting impact on our climate.

We have an obligation to be honest and open about what the scientific evidence says about the reality and real dangers of climate change.

Yes, the climate is changing.

Yes, humans are contributing to this change.

Yes, we want a strong Environmental Protection Agency that protects human health and the environment.

Yes, we want to work together to find solutions to the global threat of climate change.

Yes, we want an Administration that listens to the scientific evidence, and does not hide the truth about the consequences of pollution or climate change from its citizens.

No is not the answer. Denial and misdirection will not lead to solutions. We should work together to address and mitigate the economic consequences and social costs of our new climate reality.

Thank you. I yield back.

136

**Statement
On Behalf of the
American Road and Transportation Builders
Association**

**Submitted to the
United States House of Representatives
Committee on Science, Space and Technology
Subcommittee on Environment
and
Subcommittee on Oversight**

Hearing on At What Cost? Examining the Social Cost of Carbon

February 28, 2017

Chairman Biggs, Chairman LaHood, Ranking Member Bonamici and Ranking Member Beyer, thank you for holding this hearing on At What Cost – Examining the Social Cost of Carbon. ARTBA, now in its 115th year of service, provides federal representation for more than 6,000 members from all sectors of the U.S. transportation construction industry. ARTBA's membership includes private firms and organizations, as well as public agencies that own, plan, design, supply and construct transportation projects throughout the country. Our industry generates more than $380 billion annually in U.S. economic activity and sustains more than 3.3 million American jobs.

Because of the nature of their businesses, ARTBA members undertake a variety of activities that are subject to environmental laws and regulations. ARTBA's public sector members adopt, approve, or fund transportation plans, programs, or projects which are all subject to multiple federal regulatory requirements. ARTBA's private sector members plan, design, construct and provide supplies for federal-aid transportation improvement projects. The "Social Cost of Carbon" (SCC) has been proposed as a component of a number of regulations impacting the development and maintenance of needed transportation infrastructure facilities. Most recently, ARTBA submitted comments on August 1, 2016, to the United States Department of Transportation (U.S. DOT) objecting to the use of SCC in benefit-cost analysis for rail projects.

The SCC, which was developed in 2010 by a group of 13 federal agencies, including the U.S. DOT, is "an estimate of the monetized damages associated with an incremental increase in carbon in any given year." ARTBA has two major critiques of the SCC. First, when applied to transportation projects, the SCC is only used to ascertain the "costs" of those projects in terms of carbon emissions. SCC does not account for any carbon-related "benefits" achieved through the construction of transportation improvements which reduce congestion. EPA's own data indicates that carbon emissions from vehicles are higher when vehicles are stuck in congestion as opposed to free-flowing traffic. Thus, SCC is an incomplete or biased analysis. If there is to be a tool measuring the implication of carbon emissions from transportation improvements, it must include the benefits realized by those projects.

Further, the vagueness of the SCC presents a variety of dangers to the regulated community. The proposal's lack of specifics as to what types of "costs" are to be measured could enable project opponents to suggest an endless array of considerations which would essentially preclude new transportation improvements from being built. For example, many project opponents believe in the theory of "induced demand," which essentially states that any new road capacity will "create" new motor vehicles to occupy it. Thus, use of the SCC to account for "induced demand" would be holding transportation projects responsible for the worst-case scenario predictions of their opponents.

Additionally, the open-ended nature of the SCC could attempt to hold transportation projects responsible for emissions associated with development occurring after the project is completed. Put another way, a new road could be held accountable for emissions coming from houses and/or businesses built along the road after it is complete. Again, such a measurement would be heavily speculative at best.

In our August 2016 comments, ARTBA voiced concern over the inclusion of SCC in guidance mandated by the "Fixing America's Surface Transportation" (FAST) Act surface transportation reauthorization law. Specifically, ARTBA noted that inclusion of SCC in the guidance exceeds both the authority of the Federal Railroad Administration and the intent of the FAST Act. Nowhere in the FAST Act is the FRA (or U.S. DOT) instructed to analyze whether or not GHG emissions equate to any type of cost or benefit for rail projects (or any other transportation projects). Congress had a chance to include greenhouse gas (GHG) related measures in the FAST Act when it was deliberated in both the House and Senate and chose not to do so. In addition, the Office of Management and Budget released a November 2, 2015 Statement of Administration Policy during the FAST Act negotiations which also failed to convey an Administration recommendation that GHG analysis should be a part of the legislation's implementation.

ARTBA's comments underscore our broader concern about SCC as an analytical tool—specifically that it exceeds the authority of the federal government and it was promulgated without proper input from the regulated community. Various organizations have raised concerns over the methods used in calculating SCC and whether or not SCC has undergone an adequate notice and comment process in prior agency rulemakings. While some federal agencies may disagree with these concerns it is important that such issues be fully resolved before SCC is used in any agency guidance or regulation.

To illustrate this point, the guidance on which ARTBA commented in 2016 notes that the monetary values for carbon emissions reduction "were constructed by discounting the damages caused by its contribution to changes in the global climate from that year through the distant future." While opinions on climate change may differ amongst parties, this is hardly the type of definitive measurement that should be used in assessing the cost and/or benefits of any transportation project.

The SCC as a measurement needs to be further defined before it is used in this (or any other) guidance and/or regulation. This should be accomplished by further study and additional opportunities for participation and comment by the regulated community.

ARTBA looks forward to continuing to work with the subcommittees towards achieving a cleaner environment through efforts which strike the proper balance between proper regulatory protection and our nation's infrastructure needs.